Future of Business and Finance

The Future of Business and Finance book series features professional works aimed at defining, describing and charting the future trends in these fields. The focus is mainly on strategic directions, technological advances, challenges and solutions which may affect the way we do business tomorrow, including the future of sustainability and governance practices. Mainly written by practitioners, consultants and academic thinkers, the books are intended to spark and inform further discussions and developments.

Patrick Siegfried

Digitalisation in Mobility Service Industry

A Survey-based Expert Analysis

 Springer

Patrick Siegfried (iD)
International Management
ISM International School of Management GmbH
Frankfurt am Main, Germany

ISSN 2662-2467　　　　　　　ISSN 2662-2475　(electronic)
Future of Business and Finance
ISBN 978-3-031-07150-8　　　　ISBN 978-3-031-07151-5　(eBook)
https://doi.org/10.1007/978-3-031-07151-5

This Springer imprint is published by the registered company Springer Nature Switzerland AG
The registered company address is: Gewerbestrasse 11, 6330 Cham, Switzerland

Contents

List of Abbreviations

ADV	Autonomous driving vehicles
AMODS	Autonomous mobility on-demand services
ARS	Acquiescence response style
Bn	billion
CCM	Customer Centric Management
CEM	Customer Experience Management
cf.	confer
CRM	Customer Relationship Management
DMS	Digital Mobility Split
EMI	European Metropolitan Network Institute
ERS	Extreme response style
f.	following
GMP	German Mobility Panel
HBR	Harvard Business Review
ICT	Information and communications technology
IoT	Internet of Things
IPCC	Intergovernmental Panel on Climate Change
LOHAS	Lifestyle of health and sustainability
MaaS	Mobility as a service
Mn	million
MPT	Motorised private transport
NBO	Next best offer
ODRS	On-demand ride services
OEM	Original equipment manufacturer
p.	Page
PAYG	Pay as you go
PAYGO	Pay as you go options
POC	Point of completion
R&D	Research and development
RMV	Rhein-Main-Verkehrsverbund
SDRB	Social desirability response bias
SLA	Service level agreement
SPSS	Statistical Package for the Social Sciences

TCO	Total cost of ownership
UD	Universal design
URB	Uninformed response bias
USP	Unique selling proposition
UN	United Nations
UNDESA	United Nations Department of Economic and Social Affairs
VDV	Association of German Transport Companies

List of Figures

List of Tables

The Digitalisation in Mobility Service Industry

1

1.1 Background

> Digitization is rewriting the rules of competition, with incumbent companies most at risk of being left behind (Hirt & Willmott, 2014, p. 1).

The emergence of disruptive technologies has reshaped several industries. Even though such technologies underperform established products in the near-term disruptive innovations create (new) customer value by being cheaper, simpler or more convenient to use (cf. Christensen, 1997, p. 11). Digitalisation[1] has proven itself to possess this potential. Mostly by lowering market entry barriers and disaggregating value chains digitalisation enabled new entrants to compete rapidly with established market players at often lower costs (cf. Hirt & Willmott, 2014, p. 2). Although digital transformation is no recent phenomenon the current phase of digitisation is regarded as the most consequential. After the first two phases (i.e. the "dot com era" and the era of web 2.0), the third wave of digital transformation may have the most profound implications on an industry and its companies (cf. Evans & Forth, 2015, 3 f.). The pace at which new products and business models occur has accelerated substantially and has not been observed before (cf. Osterwalder & Pigneur, 2011, p. 9). Newmarket entries are continuously challenging established companies thereby disrupting long-successful business models (cf. Weill & Woerner, 2015, p. 27).

It might be claimed that after decades of stability the threat of disruptive innovations has reached the gates of the mobility service industry. As digitalisation gradually dissolves industry barriers the threat of newmarket entrants has eventually hit mobility providing companies.[2] In addition, the requirement to digitise own products in order to remain competitive has become a top priority of many

---

[1] Please note that the terms 'digitisation' and 'digitalisation' are utilised interchangeably.

[2] As mobility service providers or mobility providing companies firms which are involved in the production of physical assets (e.g. car manufacturers) or in the provision of services related to mobility is understood.

established companies (cf. Weill & Woerner, 2015, 27 f.). The increasing variety of mobility related services (e.g. navigation applications) offered by internet and technology companies such as Google and Apple could be regarded as a first sign of the industry's digital disruption (cf. Siegfried, 2013, p. 173). Moreover, the involvement of these companies in the development of autonomous driving vehicles (ADV) could be seen as a clear indicator of such a forthcoming transformation as well.

Nevertheless, digitalisation provides companies with an abundance of feasibilities to improve the manner business activities are conducted in current state as well. In the course of digitalisation interactions between a company and its suppliers, customers and employees are considered to improve drastically (cf. Hirt & Willmott, 2014, p. 2). The increasing availability of customer-specific data could be utilised to improve customer relationships by individualising service offerings as well as improving cross-selling opportunities. Since at present state mobility services or rather mobility systems are mostly designed according to the needs of its operators (cf. Spickermann et al., 2014, p. 215) digitisation might be utilised to shift this focus to its customers. Furthermore, the industry's digital transformation might aid in tackling several issues current mobility systems are facing (cf. Siegfried, 2014, p. 20). Current systems are suffering amongst others from capacity imbalances as well as interference with the sustainability goals of its countries, mainly due to the emission of greenhouse gases (GHG). Hence, the development of more sustainable mobility systems is sought after (cf. Cohen & Kietzmann, 2014, p. 280). One of the most ambitious goals in this regard was announced by the Finnish municipality of Helsinki which intends to make car ownership redundant by 2020 (cf. Greenfield, 2014).

1.2   Purpose of the Research

As digitalisation has altered the business environment of numerous industries it seems crucial to evaluate potential implications this technology might have on the mobility service industry. The previously mentioned blur of industry boundaries and minimisation of market entry barriers could be regarded as factors creating an impetus for a paradigm change in the industry's structure. In addition, the feasibility of improving customer relationships provided by this technological trend might call for a transformation from plain mobility management to a true customer-centric management (cf. Siegfried, 2015, p. 135). As observed in various industries before, digitalisation might lead to stronger personalisation tendencies of the offered products and services (cf. Tils et al., 2015, p. 8).

The purpose of this research is thus to explore and assess mobility in the age of digitalisation (in short: digital mobility). This assessment should include the analysis of both alterations in the relationship between a company and its customers as well as a potential shift of paradigm in the market leading companies. Moreover, a method to quantify this impact should be developed. The evaluation should be carried out from a consumer's point of view, as previous studies have already focussed on surveying business executives or experts (see e.g. Cuddihey et al., 2015). Moreover, the research should focus on the developments in Germany from a management perspective.

As a basis for analysis this research will draw upon publications in relevant academic journals, textbooks as well as practitioners' literature and consultancy reports. The main rationale behind this may be the strong practical importance of this topic. Besides, broadening the scope of publications might yield additional valuable insights since the current state of scientific research in this area is rather underdeveloped. Besides the usage of secondary literature, data gathered in an empirical research carried out in cooperation with Messe Frankfurt Exhibition GmbH will function as a second source for the analysis.

This research follows a functionalistic research process as described by Bhattacherjee (2012, pp. 20–23) and is divided into seven chapters. The first three chapters represent the exploration phase of the functionalistic paradigm. In Chap. 1, the reader has been introduced to the research's environment. Moreover, the overarching research problem was stated. Chapter 2 characterises the first half of the conducted literature review which is concerned with the definition of central terms and processes such as the satisfaction of mobility needs. In addition, Chap. 2 outlines macro-environmental factors that are regarded as key drivers and enablers of digital mobility. The main components of digital mobility systems are defined in the chapter as well. The subsequent Chap. 3 depicts the second half of the literature review and provides a rather micro-environmental perspective. In this chapter, altering requirements on mobility services as well as current and future customer centric management practices are discussed. The chapter is closed with an overview of business model archetypes and potential companies that may act as digital mobility services providers.

Chapter 4 represents the research design phase in which the overarching research problem is operationalised by dividing it into specific research questions. Besides, the selection of research method and instrument as well as the sampling strategy and the components' design is explained.

The research execution phase is portrayed by Chaps. 5 and 6. Chapter 5 describes and discusses findings drawn from the pretest and main research. In Chap. 6 these finding are utilised to develop two scenarios of future mobility markets. In addition, a model to quantify the impact of digitalisation on the mobility service industry is discussed.

The final Chap. 7 comprises a critical review of the insights drawn from this research and provides a summary of key findings. Moreover, major limitations and recommendations for further research are stated.

References

Bhattacherjee, A. (2012). *Social science research: Principles, methods, and practices.* Textbooks collection: Book 3. University of South Florida.

Christensen, C. M. (1997). *The innovator's dilemma: The innovator's dilemma. When new technologies cause great firms to fail.* Harvard Business School Press.

Cohen, B., & Kietzmann, J. (2014). Ride on!: Mobility business models for the sharing economy. *Organization & Environment, 27*(3), 279–296. https://doi.org/10.1177/1086026614546199

Cuddihey, A., Butler, S., Schneider, E., Sim, H., & Wilson, M. (2015). *The AccentureTechnology vision 2015: Public transportation: Riding the digital era.* Accenture.

Evans, P., & Forth, P. (2015). *Navigating the world of digital disruption.* The Boston Consulting Group.

Greenfield, A. (2014). *Helsinki's ambitious plan to make car ownership pointless in 10 years.* Retrieved August 20, 2021, from https://www.theguardian.com/cities/2014/jul/10/helsinki-shared-public-transport-plan-car-ownership-pointless

Hirt, M., & Willmott, P. (2014). *Strategic principles for competing in the digital age* (McKinsey Quarterly No. May 2014). McKinsey & Company.

Osterwalder, A., & Pigneur, Y. (2011). *Business Model Generation: Ein Handbuch für Visionäre, Spielveränderer und Herausforderer.* Campus Verlag.

Siegfried, P. (2013). *The service engineering concept for business.* Entrepreneurship-conference University of Lisboa, 19.-23.08.2013 (pp. 173–187).

Siegfried, P. (2014). *The importance of the service sector for the industry, teaching crossroads: 9th IPB Erasmus Week* (pp. 13–23). Bragança: Instituto Politécnico. ISBN: 978-972-745-166-1.

Siegfried, P. (2015). Die Unternehmenserfolgsfaktoren und deren kausale Zusammen-hänge, Zeitschrift Ideen- und Innovationsmanagement (pp. 131–137). Deutsches Institut für Betriebs-wirtschaft GmbH/Erich Schmidt Verlag. https://doi.org/10.37307/j.2198-3151.2015.04.04

Spickermann, A., Grienitz, V., & von der Gracht, H. (2014). Heading towards a multimodal city of the future? *Technological Forecasting and Social Change, 89,* 201–221. https://doi.org/10.1016/j.techfore.2013.08.036

Tils, G., Rehaag, R., & Glatz, A. (2015). *Carsharing – ein Beitrag zu nachhaltiger Mobilität* (Working Papers des KVF NRW No. 2). Verbraucherzentrale NRW.

Weill, P., & Woerner, S. L. (2015). Thriving in an increasingly digital ecosystem. *MIT Sloan Management Review, 56*(4), 27–34.

Mobility and Demand in the Current Literature Discussion

2.1 Mobility and Demand

Even though mobility is a term frequently used in everyday language, there is no commonly accepted or universal definition (cf. Tully & Baier, 2006, p. 30; Zierer & Zierer, 2010, p. 19). Since mobility is derived from the Latin term 'mobilitas', which signifies 'movable', it is often referred to as the movement or movability of objects such as persons, goods or information (cf. ARL, 2005, p. 654; Ahrend et al., 2013, p. 2; Zierer & Zierer, 2010, p. 19). It could be argued that due to this lack of a common definition the term is utilised differently in different academic disciplines, even within the social sciences. Therefore, several attempts to define a broader concept of mobility can be found in the body of knowledge. Already in the late 1990s Cerwenka (1999) developed a categorisation scheme to differentiate the terminology found in the respective literature.

As depicted in figure 'Dimensions of Mobility' (cf. Cerwenka 1999, p. 351) what is generally referred to as mobility can be divided into three main elements, which are not fully independent from each other. *Social mobility* refers to the movement between social classes or the change in one's social situation and status (cf. Zierer & Zierer, 2010, p. 21). Movements in this dimension are rather time-dependent and—in theory—do not require movements in space (cf. Ciftci et al., 2022, p. 57). Moreover, it limits an individual in the number of options to realise spatial mobility (i.e. cost constraints). Ammoser and Hoppe (2006, p. 9) describe *mental mobility* as the ability to think flexibly, evaluate potential action alternatives as well as expand one's 'space of possibilities' abstractly. It is moreover seen as a prerequisite for spatial mobility by the authors. In addition, it may be stated that mental mobility capabilities correlate with constraints in social mobility. *Spatial (physical) mobility* describes the movement of individuals or households between elements within a defined system (cf. Windzio, 2013, p. 664). This dimension on the other hand influences a person's social mobility competencies in terms of required mobility for occupying a specific job. Cerwenka (1999) further divides spatial mobility into *migration mobility* and *traffic mobility*. While migration mobility describes a rather

© The Author(s), under exclusive license to Springer Nature Switzerland AG 2022
P. Siegfried, *Digitalisation in Mobility Service Industry*, Future of Business and Finance, https://doi.org/10.1007/978-3-031-07151-5_2

permanent movement from one defined place in a system to another, traffic mobility represents regular movements between several places.

Although traffic mobility is often used synonymously for circular mobility or everyday mobility (cf. e.g. ARL, 2005, p. 655), Windzio (2013, p. 663) argues that there are considerable differences in the meaning of those two terms. By discussing relevant definitions, Windzio (2013, p. 663) points out that circular mobility in a narrow sense solely describes movements that are traced back to work commuting, disregarding movements entailed by leisure activities such as grocery shopping or sports activities, which are incorporated in the definition of everyday mobility (cf. Lempp & Siegfried, 2021, p. 47).

Tully and Baier (2006) pursue a similar approach to categorise the existing terminology. Likewise to Cerwenka (1999), the authors differentiate between social and spatial mobility as well as relate spatial mobility to the duration of the movement (i.e. migration or everyday mobility). Notwithstanding, the existence of mental mobility (in a narrow sense) in the overall scheme is neglected. Instead, *informational mobility* is considered to be a main dimension. Informational mobility in this context refers to an exchange of information without considerable time delay or spatial movement through chemical, electrical or other processes (cf. Tully & Baier, 2006, p. 33). Nonetheless, it might be claimed that informational mobility contains what is referred to as mental mobility by Ammoser and Hoppe (2006, p. 9). Tully and Baier (2006, p. 33) state that every individual may act as a sender or recipient of information. Accordingly, if the sender and recipient are the same person this information transmission might be referred to as thoughts. Hence, it could be assumed that these thought processes also contain evaluations of potential action alternatives and the expansion of one's 'space of possibilities' mentioned by Ammoser and Hoppe (2006, p. 9).

In accordance with Cerwenka (1999), Tully and Baier (2006) acknowledge the interrelation of all mobility components as well. Furthermore, the authors claim that a fourth dimension should be considered as a main element of mobility: beaming (i.e. the movement in space without considerable movement in time). Nevertheless, from the current perspective on technological progress and feasibility of movements via beaming, the dimension is neglected.

Considering the aim of this research it seems appropriate to develop and utilise a hybrid form of the concepts of Cerwenka (1999) and Tully and Baier (2006) to serve as a general framework. While the structure is derived from Cerwenka (1999) the defined mental mobility dimension will be replaced by the informational mobility dimension proposed by Tully and Baier (2006). The main reason may be that informational mobility is not restricted to the exchange of an individual's thoughts, but also incorporates the exchange of digital or virtual information (cf. Tully & Baier, 2006, p. 33). In addition, mental mobility is highly dependent on information availability (cf. Ammoser & Hoppe, 2006, p. 9) which is—in the view of the author—better reflected in the definition of informational mobility.

The focus of this paper should be on the dimension of spatial mobility, in particular on everyday mobility as defined by Windzio (2013, p. 663). Moreover, short-time but long-range movements such as tourism as described by Tully and

Baier (2006, p. 31) will be regarded as a relevant subcategory of everyday mobility as they could be considered as part of leisure activities.

2.1.1 Need for Mobility

It may be claimed that every human movement can be traced back to the satisfaction of a specific need. The need for mobility in a general context is defined as the desire for overcoming spatial distances in order to satisfy a specific need or carry out certain activities (cf. Groß & Freyer, 2010, p. 2; Ahrend et al., 2013, p. 17). Eriksson (2011, p. 3) specifies these needs into biological needs, social obligations and personal desires which often do not take place simultaneously and therefore may cause one or several trips.

As demonstrated in the 'The Driving Forces of Mobility Needs' (cf. Eriksson, 2011, p. 4) above the occurrence of a need (biological, social, personal) creates a demand on the performance of an activity. This activity on the other hand produces a need for travel or rather the need for mobility. This need is influenced by the spatial organisation of the environment as well, since this factor might have implications on the distance which has to be overcome. The selection of destination, departure time and travel mode is again influenced by the availability of modes in the individual's environment (see Sect. 2.1.2).

Bartz (2015, p. 32) points out that the need for mobility can be categorised into two overarching motives. The first motive describes the satisfaction of one's mobility need by the actual usage of a transport mean. In this case mobility is perceived as a primary need. The second main motive could be defined as the utilisation of transport modes as means to an end. Mobility thereby is perceived as a secondary need which is necessary to overcome spatial distances in order to satisfy the individual's primary needs such as job commuting, grocery shopping or social relations (i.e. the need for mobility in a general context). Bartz (2015, p. 33) underlines that the selection of transport modes is of greater importance when mobility is considered to be a primary need, which will be discussed in the following chapter.

2.1.2 The Process of Travel Mode Selection

The selection of travel modes is a complex construct that has received significant attention from academic researchers of different disciplines. The process consists of a three-level decision-making that considers subjective and objective decision factors (cf. Delatte et al., 2014, p. 8). Delatte et al. (2014, p. 9) claim that the final decision on an individual's travel mode mix is mainly influenced by the first two decision levels, which are based on objective factors. The underlying reason might be that those factors have a stronger limitation character than subjective factors. On the third level, subjective factors are taken into account to finally decide upon a mode of transport. Bartz (2015, p. 40) also acknowledges that the selection of transport means is a multicausal process that is dependent on structural and individual parameters. EMI (2012, p. 42) support the two previous views by claiming that travel choices do not occur in a vacuum, but rather in a complex web of constraints, life choices and conditions under which those decisions are made. Furthermore, these factors and requirements are subject to constant change as life events continuously shape and alter a person's mobility needs (cf. Scheiner, 2007, p. 163).

As objective or structural factors Delatte et al. (2014, p. 9) and Bartz (2015, p. 39) identify the availability of different modes of transport as well as spatial and settlement structures. Furthermore, it may be argued that individual socio-economic factors limit a person in the selection of transport modes. Groß and Freyer (2010, p. 2) mention the possession of a driver's licence in order to be eligible to drive mortised vehicles. EMI (2012, p. 41) and Otto (2010, 1 f.) state that age, gender and income are variables to be considered.

Several studies have been dedicated to determining the subjective factors which shape the overall decision-making process. To achieve this, most studies relied upon psychological theories such as the theory of planned behaviour or the rational choice theory. The findings indicate that the travel mode choice is mainly founded on reasoned decisions based on a person's attitudes and values (cf. EMI, 2012, p. 42). Delatte et al. (2014, p. 9) in particular highlight an individual's subjective perception of its environment. By reviewing related studies, the authors demonstrate that for instance alternatives to the individual car usage are often perceived to be less attractive than their actual attractiveness or suitability for the purpose of travel (cf. Siegfried, 2021, p. 124). Bartz (2015, p. 29) supports this viewpoint by identifying expected service performance as a criterion for the transport mode selection. Groß and Freyer (2010, p. 10) and Musti and Kockelman (2011, p. 708) mention personal needs, attitudes, preferences as well as one's lifestyle to be determining subjective factors. These statements comply with the results of Vredin Johansson et al. (2006, p. 518) that draw upon findings of their empirical research which conclude that attitudes and personality traits have a great influence on mode choice. According to EMI (2012, p. 41) to completely understand and evaluate the variances in personal travel patterns, a person's social network and the physical distance between the contacts has to be considered as well.

Delatte et al. (2014, p. 10) summarise the travel mode selection process as a function consisting of an instrumental motive (means to an end), social expressive motive (symbolic meaning) as well as an intrinsic or emotional motive (end in itself). A non-exhaustive list of own modified feasible factors is provided in Table 2.1.

2.1.2.1 The Influence of Habits on Travel Mode Selection

A second major research stream claims that the transport mode selection is mostly based on habitual processes. A habit refers to a more or less fixed behaviour in a certain situation which has been adopted through continuous repetition (cf. Andrews, 1903, p. 122) that is often formed subconsciously. Studies have concluded that the mode choice follows standardised routines for certain routes or the selection of certain transport modes for specific purposes without the consideration of alternatives (cf. EMI, 2012, p. 42; Bartz, 2015, p. 42; Delatte et al., 2014,

Table 2.1 Motives of travel mode selection (Source: Own depiction)

Instrumental Topics	Social expressive Topics	Intrinsic Topics
• Spatial availability • Accessibility of transport mean • Ubiquitous accessibility • Comfort/convenience • Safety and security • Mobility at destination • Handicapped accessible	• Communication of status and prestige • Social distinction • Free choice on level of privacy • Satisfaction of social norms or expectations of others	• Freedom of choice • Sense of independence • Flexibility and spontaneity • Pleasure of driving

p. 9). Otto (2010, p. 1) argues that attitudes alone are insufficient to describe an individual's behaviour in the travel mode selection process and, hence, the influence of routines and habits should be considered to a much greater extent. The author justifies his assumption with the allegation that habits are deeply rooted in a person's behaviour as they assist in the simplification of everyday tasks.

One of the main rationales behind the application of habits in transport mode selection is the compensation of a decision-making on a too complex, unclear and potentially biased basis of information. Delatte et al. (2014, p. 9) underline the information overload an individual is facing and therefore opts for the application of habits as an effective strategy to reduce cognitive effort. In accordance with Delatte et al. (2014), Bartz (2015, p. 42) points out that the collection and processing of all decision relevant data are impossible to achieve. The author argues that due to this lack of capabilities the decision on transport modes is biased to a varying degree between different vehicles. Accordingly, the selection is therefore rather based on the attractiveness of a mode as on the suitability for the purpose of travel.

2.1.2.2 The Role of Information in Travel Mode Selection

As it could be concluded from the previous paragraphs information has a significant impact on the travel mode selection. The provision of decision relevant information may be seen as vital for the selection process. Otto (2010, p. 2) sees the availability of information and knowledge as crucial preconditions for behavioural changes in transport mode choice. In their research, Grotenhuis et al. (2007, 34 f.) analysed the travellers' information requirements in the three stages of a journey (i.e. pre-trip, wayside, and on-board). The research revealed that travellers in general perceive information to be valuable if time (e.g. search and travel time) or effort (e.g. physical and mental effort) savings could be achieved. This could imply that if relevant information to reduce time and effort could be gathered in a convenient manner, current routines in the selection process can be altered.

Notwithstanding, some studies suggest that even though sufficient information might be obtained the travel mode choice is still biased or information is neglected. Lenz (2011, p. 613) claims that, since repetitive journeys such as work commuting are based on habits and routines, the inclusion of new information into the decision-making process is hindered. Delatte et al. (2014, p. 9) support this opinion and additionally underline the application of subjective filters on available information to remove information considered to be inadequate to a person's attitudes and beliefs.

2.1.3   Satisfaction of Mobility Needs

The satisfaction of (physical) mobility needs is achieved through participation in traffic. The relationship between mobility and traffic is often misunderstood and the two terms are frequently confused and utilised as synonyms (cf. Zierer & Zierer, 2010, p. 23). Therefore, it seems crucial to briefly review the connection between traffic and mobility. Traffic is defined as the actual change of places of persons, goods or information (cf. Huber & Laverentz, 2012, p. 196) while (spatial) mobility rather describes the potential of objects to move between places. Becker (2011, p. 81) states that mobility is rather concerned with aspects regarding the demand for spatial movement and hence represents the demand side; traffic on the other hand comprises all aspects of the realisation of spatial movement and thus represents the instrumental side of the construct. Henceforth, traffic could also be seen as realised and quantified spatial mobility (cf. Deffner & Götz, 2010, p. 11; Tully & Baier, 2006, p. 34).

To participate in traffic the literature generally differentiates into two overarching categories: private and public transport. *Private transport* comprises all transport modes which are utilised by solely one or a limited amount of persons and includes amongst others bicycles and walking. Private transport is mostly driven and triggered by the using individual's purposes for travel (cf. Ammoser & Hoppe, 2006, p. 5). The selection of transport mean, time of departure and the destination of the journey are determined by the user (cf. Henkel et al., 2015, p. 3). A special form of private transport is motorised private transport (MPT) which consists of cars, motor bikes and other motorised vehicles, and is the most prominent mode of transport in many countries. This might be due to the high degree of flexibility in terms of time of departure and route selection as well as the feasibility to overcome large spatial distances rather quickly. It could also be claimed that vehicles used in private transport are—at present state—mostly owned by the using individual. *Public transport* on the other hand consists of transport modes which are used by many persons such as trains, metros and busses. In addition, public transport modes are usually based on scheduled departures and assigned to fixed routes, thus not driven by an individual's purpose for travel (cf. Henkel et al., 2015, p. 3). Moreover, those vehicles are usually part of a transport operator's service provision and not owned by the using individuals.

The quantification of realised mobility is depicted in a country's or city's modal split. The *modal split* indicates the share of the respective transport modes based on a determined parameter such as passenger kilometres, number of trips or average hours of usage per week (cf. Firnkorn, 2012, p. 1662). Bartz (2015, p. 31) argues that the quantification of mobility should rather be differentiated on an individual's and aggregated level. The author claims that a person's share of utilised transport modes is quantified as its *mode mix* whereas the modal split measures the share of movements of transport modes within a geographical region. Furthermore, it might be stated that the quantification of mobility does not follow a one-to-one conversion, but rather depends on the usage of transport types. An individual might opt for the satisfaction of his mobility need by the usage of a train. This may produce additional passenger kilometres, however, no additional vehicle has to be introduced into the transport system (at least in short-term). This 'mobility quantification', however, depends on the variable the modal split is measured on. Therefore, it may be claimed that traffic produces mobility; however, mobility does not necessarily produce traffic. This might also hold true if one considers the fact that not all objects that are considered to be mobile make use of their mobility potential.

2.2   Digital Mobility

In line with the term mobility, digital mobility lacks an overall and commonly accepted definition. Moreover, there is no consensus in the currently available literature on the exact terminology. In the literature review conducted for this research four major terminology streams were identified that may relate to an identical idea of a novel mobility paradigm: smart mobility, intelligent mobility, mobility as a service (MaaS), and future mobility.

Wolter (2012, p. 528) defines *smart mobility* as a service that allows the realisation of energy efficient, low emission, safe, comfortable and cost-efficient mobility that is used intelligently by all traffic system users with the support of information and communication technologies. Lauwers and Papa (2015, p. 3) rather associate smart mobility with recent developments and the altering conditions in the transport sector. With reference, the authors summarise smart mobility as the major changes caused by the introduction of new technologies, products and services that lead to a change in users' and providers' expectations and opportunities as well as their impacts on transport systems.

The term *intelligent mobility* is proposed amongst others by the Automotive Council UK (2011) and the National IT Summit (2015). According to the Automotive Council UK (2011, p. 6) the term refers to a concept that should integrate all kinds of transport modes rather than their standalone provision thereby heavily relying on electronic devices and communications systems to optimise the performance of the overall system. The National IT Summit (2015, p. 6) provides a similar definition, however, additionally highlights the individual, cost-efficient and environmentally friendly service provision as well as the potential to jointly utilise or rather share offered services.

As underlined by the Finnish Transport Agency (2015, p. 9), the terminology around the concept of *mobility as a service* experiences a broad range of interpretations of both the term itself as well as its scope. Following the authors' rather broad interpretation MaaS is a bundled market offering for consumers that includes at least one mobility-related service to ensure easy and reliable travelling. Kamargianni et al. (2015, p. 12) understand MaaS as the procurement of mobility services according to consumer needs that enable users to design seamless door-to-door mobility chains without buying or possessing the required mean(s) of transport. In line with the previously outlined definitions, the authors see ICT as a central component of the concept.

The literature review also identified several publications which do not define a specific term, but rather describe *future mobility* systems. A common characteristic of this terminology stream is the viewpoint on those systems as an ecosystem of mobility and related services. Corwin et al. (2015, p. 16) argue that, similar to other industries, the mobility market is undergoing a metamorphosis caused by the increasing implementation of digital technologies. This transformation may lead to the establishment of an environment of highly interconnected and specialised businesses cooperating with one another, across industry boundaries, to achieve a more fluid ecosystem to serve users' mobility needs (cf. Corwin et al., 2015, p. 16). Cuddihey et al. (2015, p. 3) describe the mobility system of the 'We Economy' as a rich and interconnected ecosystem that consists of third parties, partner agencies and vendors that collaborate to provide services in an entirely different manner. Fishman (2012, p. 9) advocates for a similar transformation. The author states that future transport systems might consist of a combination of transport modes, services and technologies. Further, Fishman (2012, pp. 11–15) claims that this ecosystem shares five major characteristics: massively networked system elements, user centricity, integration of different services, dynamically priced as well as driven by the cooperation of public, private and non-profit entities. Capgemini (2013, pp. 6–9) shares the view of Fishman (2012). In addition, Capgemini (2013) considers the sustainability of such a system as a main characteristic.

As mentioned before it might be claimed that even though there is no consensus on the exact terminology thus far, to a certain extent the different terminology streams describe a similar concept. Notwithstanding, none of the above-provided definitions satisfies the requirements to be referred to as comprehensive in terms of incorporation of all characteristics associated with digital mobility in the context of this research. Hence, it was decided to develop a more precise definition.

By comparing the previously outlined definitions it could be stated that digital mobility possesses five major characteristics: sustainability[1] (cf. Wolter, 2012; National IT Summit, 2015; Capgemini, 2013), user centricity (cf. National IT Summit, 2015; Finnish Transport Agency, 2015; Kamargianni et al., 2015; Fishman, 2012; Capgemini, 2013), integration of all traffic carriers (cf. Automotive Council UK, 2011; National IT Summit, 2015; Kamargianni et al., 2015; Corwin et al., 2015; Cuddihey et al., 2015; Fishman, 2012; Capgemini, 2013), integration of non-mobility services (cf. Corwin et al., 2015; Cuddihey et al., 2015; Fishman, 2012; Capgemini, 2013) as well as steered and supported by ICT (cf. Wolter, 2012; Lauwers & Papa, 2015; Automotive Council UK, 2011; National IT Summit, 2015; Kamargianni et al., 2015; Corwin et al., 2015; Cuddihey et al., 2015; Fishman, 2012; Capgemini, 2013).

---

[1] Sustainability is interpreted according to the three bottom line approach of Elkington (1997, p. 73 f.).

Hence, digital mobility in the context of this research should be defined as an ecosystem of fully integrable and sustainable mobility and third-party services in which users are able to design door-to-door mobility chains—via electronic devices and ICT—according to their specific needs and without the necessity to possess any required mean of transport.

2.3 Enablers and Drivers of Digital Mobility

For many years studies have been discussing the so-called 'peak travel' or 'peak car' phenomenon. The phenomenon describes the specifically in Western Europe and Northern America stagnating and slowly declining demand for automobile travel. Amongst others, Germany was one of the first countries in which this phenomenon occurred (cf. Kuhnimhof et al., 2012a, p. 444). Kuhnimhof et al. (2012b, p. 64) even claim that after the United Kingdom Germany is the country in which this trend is observed most significantly. The drivers of peak travel have been intensively discussed. There is a consensus in the respective literature that the phenomenon is mainly driven by socio-economic and sociocultural shifts in the world's population which amongst others includes the growing and aging population, increasing urbanisation, the reduction of disposable income, growing sustainability concerns and the devaluation of the private car as a status symbol.

Metz (2013, 255 f.) claims that humankind has begun to enter the fourth era of travel. He states that after the first three eras (i.e. the spread of the human population during the Stone Age, the Neolithic Revolution in the Middle East 10,000 years ago, and the invention of the railway in the early nineteenth century) the fourth era will be mainly characterised by the adaption of current transport systems to the socio-economic shifts in the population and the increasing influence of new technologies. Metz (2013, 255 f.) categorises the peak car phenomenon as a result of a state of transition from the third into the fourth era. These changes might demonstrate strong similarities with developments in the manufacturing sector where a transition into a fourth era is discussed as well, the so-called industry 4.0'. One major component of this new era is the increasing digitisation and interconnection or rather interaction between objects and humans. Moreover, industry 4.0 aims to enable manufacturers to realise batch sizes close to one, which creates the feasibility for true mass customisation. It may be stated that the third era of travel has to undergo similar changes in order to cope with the altering macro-environmental factors and growing demand for personalisation, thus producing demand for digital mobility to enter the fourth era.

Nonetheless, the construct of drivers and enablers around digital mobility and peak travel may be seen as highly complex. Moreover, a clear distinction between enablers and drivers might not always be feasible since many factors are strongly interrelated and therefore influence one another on many levels. The rationalisation of cars, for instance, might be a driver for a new mobility paradigm, however, simultaneously enables mobility providers to successfully introduce new mobility services which found the basis of digital mobility systems. Figure 2.1 describes these

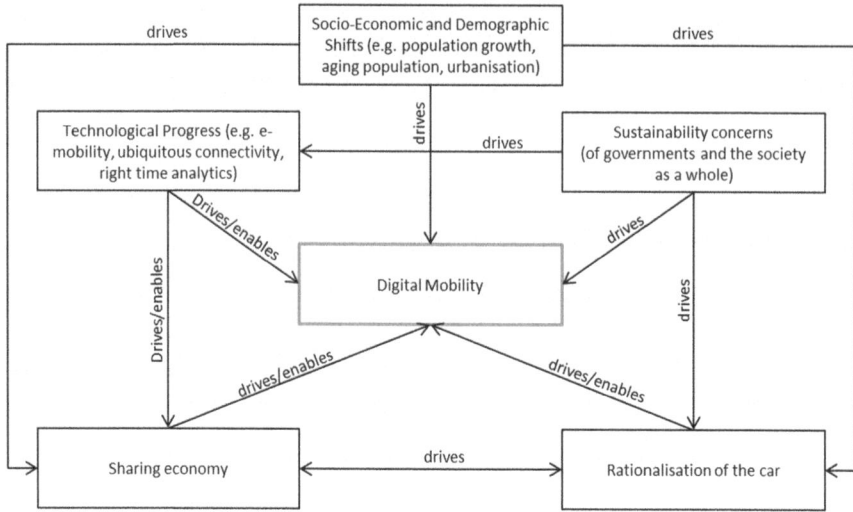

Fig. 2.1 Enablers and drivers of digital mobility (Source: Own depiction)

partly ambivalent relationships. Furthermore, it appears necessary to analyse factors driving and enabling digital mobility in greater detail. The following paragraphs will scrutinise the main factors mentioned in the respective literature associated with this potential shift of paradigm and how they impact the demand for digital mobility. Special attention will be paid to the developments related to Germany. It should be mentioned that due to the factors' complex interrelations the order of presentation in this chapter should not be seen as causal or comprehensive.

2.3.1 Technological Progress

Unarguably, technological progress may be the most significant enabler of digital mobility. In particular, advances in ICT and vehicle technology form the foundation for the feasibility of such a new mobility paradigm. The utilisation of the Internet and mobile devices has increased significantly over the last decade. In Germany, around 79.5% of the population uses the Internet occasionally, while 63.1% are daily users (cf. ARD/ZDF, 2016). Mobile Internet services are used by 55% and 23%, respectively (cf. ARD/ZDF, 2016). In addition, data transmission and storage technologies are continuously increasing their processing capabilities. This may lead to lower prices and a higher affordability for a larger share of people and thus to a state of almost ubiquitous connectivity. According to Rammler and Sauter-Servaes (2013, p. 27), the increasing popularity of smartphones combined with the steadily decreasing costs for data volumes are a crucial factor for the realisation of new mobility services.

As noted by Heikkilä (2014, p. 31), modern ICT facilitates the integration of services and people. In recent years the Internet of Things (IoT) has received wide attention. The IoT refers to interconnection of everyday objects which are equipped with ubiquitous intelligence with the aim to enhance the interaction between humans and devices as well as between the devices itself (Xia et al., 2012, p. 1101). It may be stated that the establishment of such a system is the foundation of digital mobility ecosystems. In digital mobility systems data from several sources such as the physical infrastructure, vehicles or services is continuously collected and analysed in order to enhance the performance of the overall system (cf. Buscher et al., 2014, 23 f.). In addition, the data might be required to provide users with real-time information on departures as well as to allocate or rather reserve vehicles or seats. In this regard, big data and right time analytics might play a major role in order to process a large amount of data these systems will produce. Such analytical techniques on the one hand could enable traffic control centres to decrease their response time as well as the support of user decisions on a moment-to-moment basis (cf. Fishman, 2012, p. 8) in accordance with the user's specific requirements or preferences.

In the field of vehicle technology, the development and enhancement of alternative engines such as electric drive technology are considered a major enabler for digital mobility. This might be traced back to the fact that such alternatives are less dependent on finite resources such as crude oil. Moreover, electric cars demonstrate significantly lower greenhouse gas emission rates, and therefore should be seen as a more sustainable alternative to regular cars with combustion engines. This might be especially crucial when the ongoing urbanisation is considered (see Sect. 2.3.2). Another major development could be seen in automation technology. The degree of automation may be divided into five categories reaching from no automation (level 0), over partially automated vehicles in terms of the takeover of certain assistive tasks, to full automation (level 4) where the driver is not required to supervise the operation of the vehicle (cf. DuPuis et al., 2015, p. 19). It could be claimed that the development of fully automated vehicles is strongly dependent on the progress in the field of ICT, especially in the establishment of the IoT.

Incorporation of all major technological enablers can be seen in the development of so-called smart cars. It is frequently argued that cars might undergo a similar transformation as it was observed in the development from the regular mobile phone to the smartphone (cf. e.g. Fishman, 2012, p. 8; Kunze, 2015, p. 7). Smart cars might not only take over the control over the operation of itself, but will also be able to communicate with other entities in the IoT (cf. Kunze, 2015, p. 7). Moreover, the user might be able to access social media applications as well as other communication and entertainment services.

Nevertheless, as pointed out by Siemens AG (2015, p. 26) there might not only be a technological solution for the current challenges transport systems are facing. Accordingly, electro mobility for example may solve the issue of pollution, however, does not tackle congestion in general. Therefore, it could be stated that profound management knowledge in the operation of transport systems is required as well.

2.3.2 Socio-economic and Demographic Shifts

Over the past decade, significant socio-economic and demographic shifts in the world's population have been identified. These developments are frequently mentioned as major drivers for change, not only in the mobility industry. The first major trend may be the steadily growing and aging world population. Projections provided by the United Nations (UN) suggest that by 2030 the world's population will account for around 8.5 billion with a median age of 33.1 years (UNDESA, 2015, p. 3). For Germany on the other hand a decline has been predicted by various institutions (e.g. UNDESA, 2015, p. 323; Federal Statistical Office, 2009, p. 12). Nevertheless, especially Western European countries are strongly affected by the ageing population phenomenon. With a median age of 48.6 years by 2030, the German population will be 1.5 times older than the world average, for instance. A second major trend that might also affect countries with a declining population is the rising urbanisation rate. Currently, the world's urbanisation rate is around 54% (UNDESA, 2015, p. 20). In Germany, a share of 75% of the country's population is living in urban areas. Prognoses again provided by the UN predict an increase in the world's urbanisation rate by 12% to 66% by 2050; for Germany, an urbanisation rate of 83% is estimated (UNDESA, 2015, pp. 20–23).

The increasing urbanisation may impose several challenges on current urban transport systems. As cities gain in the number of inhabitants growth in traffic should be anticipated as well. Nevertheless, infrastructural expansions are impossible in many cases due to the limited availability of space (cf. Bratzel, 2014, p. 102) and financial resources (cf. Ahrens et al., 2011, p. 15). Hence, an increasing level of congestion could be expected which has significant impact on the quality of living in urban areas. Besides the reduction of efficiency and effectiveness of the overall transport system, rising traffic levels could also have negative impacts on the health of a city's inhabitants. An increase in traffic might lead to an incline in traffic-related noise and a higher likelihood of accidents. As mentioned by Cullinane and Edwards (2010, p. 37), the constant exposure to traffic-related noise could cause psychological and physiological health issues. Furthermore, the ageing population phenomenon calls for an adaptation of the current urban transport system setup to the specific requirements of this demographic which will represent a significant proportion of Germany's overall population. Ahrens et al. (2011, p. 21) also raise the importance on the increasing car affinity of pensioners which has amplified substantially over the last decade. This might produce additional traffic and thus congestion in urban transport systems.

Another considerable shift was observed in the disposable income. The imfo (2011, p. 20) points out that the average relative income in Germany has declined in the period from 2002 to 2008. Kuhnimhof et al. (2012a, p. 448) argue that this development can be mostly traced back to a shift in the biographies of the younger generations. With reference, the authors state that the share of young adults enrolling in university programmes for instance increased tremendously. This reduces the number of persons participating in a country's workforce, and hence the average income. Thus, it may be argued that future mobility systems have to take the decline in income and the entailed reduction of mobility budget of its users into account. Austin and Zielinski (n.d., p. 22) also advocate for the importance of a more affordable transport systems to meet the requirements of altering socio-economic factors.

2.3.3 The Altering Role of Private Cars

It could be claimed that the future role of private cars is the most controversially discussed driver of digital mobility. While various empirical studies concluded that the attractiveness of an own car is declining amongst younger generations (cf. Witzke, 2016, p. 23; Bratzel, 2014, p. 95; Ahrens et al., 2011, p. 21) other studies resulted in the opposite. A recent representative research for Germany conducted by Cornet et al. (2012, p. 10) for instance found that around 78% of young adults between 18 and 24 years strongly agree on the intention to own a private car in 10 years.

In order to evaluate the direction of development in Germany, it seems crucial to review the main rationales behind this declining attractiveness advocated by respective studies. One main reason may be the rationalisation or rather 'de-emotionalisation' of cars. Bratzel (2014, p. 95) summarises findings from several studies conducted between 2008 and 2011 which concluded that around one-fourth of the participants see cars as a mere means to an end. The author further notices that young adults between 18 and 25 years are less likely to reduce spending in other areas for the sake of owning a car. Rammler and Sauter-Servaes (2013, p. 25) also state that the importance to possess a car is declining significantly and is increasingly displaced by other objects. Ebel et al. (2014, p. 540) agree with the previous views. Moreover, the authors underline the societal change concerning the car's role as a status symbol, since it is progressively accepted to purchase small and inexpensive cars instead of luxury vehicles.

Another major factor that might be strongly interrelated with the growing 'de-emotionalisation' is the rise in costs associated with car ownership in combination with the improvement of public transport availability. Younger adults are less willing to bear the total cost of ownership (TCO) of private cars. The previously outlined socio-economic shifts also lead to a reduction in disposable income and thus to constraints in mode choice which might also disfavour car usage or ownership. The parking situation in city centres, as well as the risk of congestion during peak hours could, additionally contribute to a less attractive perception. There may be some empirical evidence that support these assumptions. Bratzel (2014, 95 f.) reviewed the developments in urban areas between 2002 and 2008. He concludes that while the average mobility in terms of average trip distance of young adults increased the car usage decreased by 5%. Especially low utilisation and ownership were found in cities with more than 100,000 inhabitants where car usage declined by 12% during the examination. Deffner and Götz (2010, p. 14) specifically highlight the impact of increasing public transport service provision in urban areas and the decline of car usage or rather the acquisition of driver's licences amongst the 18 and 24-year olds.

Nonetheless, recent data from the German Mobility Panel (GMP) suggest that the German car fleet is still expanding, yet with a smaller growth rate (cf. Streit et al., 2015, p. 18). In addition, the research revealed that simultaneously the fleet age increased. The share of cars older than 10 years account for around 43% of the total fleet (cf. Streit et al., 2015, p. 18). These developments could at least partly correlate with two socio-economic trends. The steady increase of cars may be moderately traced back to the increasing car affinity of pensioners which has grown over the past years. The ageing fleet on the other hand might imply a slight confirmation of a car's loss of status. As the car faces rationalisation less importance is placed on possessing the most recent model. Furthermore, it could be stated that specifically in rural areas the dependence on private cars is stronger than in urban areas due to less extensive public transport services or other alternatives. Thus, as young adults reach the age for acquiring a driver's license more cars may be required to achieve a certain level of mobility. This might also explain the strong agreement on future car ownership in research conducted by Cornet et al. (2012). To verify this assumption, however, the data would have had been made available in a different granularity. Even though the authors mention the collection of community size in their survey (cf. Cornet et al., 2012, p. 11) the data was not published.

Nevertheless, it should be stated that the role of cars as a symbol of status is changing. Those changes might be marginal at present state but still have the potential to evolve into larger trends. By drawing upon the findings of Deffner and Götz (2010) as well as Bratzel (2014) it could be claimed that if the service coverage of public transport and alternative services further increases the dependence on cars—and hence car ownership—in rural areas will decline. Notwithstanding, the transformation of cars into completely shared objects should be seen as critical. There might still be a demand for private cars. This, however, does not interfere with the increasing rationalisation since the demand for future car ownership could be driven by the demand for privacy.

2.3.4   The Sharing Economy

The private car might not be the sole object which is increasingly rationalised by younger generations. The so-called sharing economy has recently gained momentum. As starting point of this new consumption paradigm, Cohen and Kietzmann (2014, p. 279) name the need for a reduction in private spending on consumption after the global economic recession in 2008 where people began to offer and share underutilised resources. The lack of financial means (see Sect. 2.3.2) combined with increasing sustainability concerns of younger adults (see Sect. 2.3.5) are seen as key components of the current success of sharing services (cf. Goudin, 2016, 7 f.; Cohen & Kietzmann, 2014, p. 280). In 2013, the global market for sharing services was valued at $26 billion and is expected to develop into a $110 billion revenue market over the next years (cf. Cannon & Summers, 2014). Goudin (2016, p. 8) claims that the annual growth rate of the sharing economy is prognosticated to 25%, in some sectors even to 65% by 2025.

It could be claimed that in the beginning sharing of resources was mainly conducted on a peer-to-peer basis, however, experienced strong commercialisation in recent years. Progressively, established as well as new start-up companies are entering the market for shared services. Two examples of established companies may be BMW and German Railways which introduced car-sharing services in Germany. On the other hand, start-ups such as Uber are on the edge to revolutionise the taxi service market[2] by offering a platform to pair drivers with customers without possessing any assets. According to Zobrist and Grampp (2015, p. 3), the investment into companies offering shared services is steadily increasing and even surpassing investments in social media start-ups. With reference, the investment in 2015 equalled around $12 billion globally. Table 2.2 provides a non-exhaustive overview of several sectors in which such services are currently offered.

As pointed out by Goudin (2016, p. 7), there is no certainty to what extent or rather how disruptive the impact of this new consumption paradigm on the economic landscape will be. Nonetheless, it may be stated that the increasing participation of traditional companies in this domain demonstrates the vast potential the sharing economy is rewarded by the 'established economy'. Recent figures from Germany also indicate a further expansion and acceptance of sharing services as an alternative to the conventional approach. A representative research conducted by PwC (2015, pp. 5–7) found out that 46% of the total German population has at least once used a shared service in the previous 2 years (2013–2015); in the quartile of 18–29 years olds, this share rises to 88%. PwC (2015, p. 6) also found out that entertainment services (33%), followed by consumer goods (31%) and automobile and transport services (28%) are the most utilised sharing services in Germany. Another representative research by TNS Emnid (2015, p. 20) seems to prognosticate a further commercialisation or rather the requirement for stronger commercialisation. Accordingly, 62% of the survey participants prefer the usage of services offered by companies rather than private persons.

2.3.5   Sustainability Concerns

According to the German Umweltbundesamt (2015, p. 16) 24% of the CO_2 emissions that can be traced back to private consumption are caused by mobility activities. Of these, 75% are directly related to motorised private transport. The Intergovernmental Panel on Climate Change (IPCC) concluded that, from a

---

[2]It should be mentioned that Uber is still facing strong regulations when entering new markets.

Economic sector	Companies/Services
Hospitality and dining	• CouchSurfing • Airbnb • Feastly • LeftoverSwap • Housetrip
Automotive and transport	• RelayRides • Hitch • Uber • Lyft • Getaround • Sidecar • Car2Go • Moovel • ParkU • Shared Parking
Retail and consumer goods	• Neighborgoods • SnapGoods • Poshmark • Tradesy • Kleiderkorb
Media and entertainment	• Amazon Prime • Wix • Spotify • Netflix • SoundCloud • Earbits
Financial services	• Cashare • C-crowd • Kickstarter • Bondora

Table 2.2 Sharing economy sectors and operating companies

Source: Own depiction

scientific point of view, there is no doubt that the main cause of climate change is human behaviour (cf. Cullinane & Edwards, 2010, p. 37). The discussion around a more sustainable society has received significant attention over the last decades. It may be stated that due to this omnipresence of the topic stronger measures and initiatives to reduce the impact of human behaviour on the environment can be observed by both the government and the society as a whole.

Sustainability could be seen as a national objective for several years. In Article 20a of the German Constitution, the protection of the environment as well as sustainable development have been manifested. In addition, the so-called Kyoto Protocol which became effective in 2005 obliged countries that ratified the protocol to substantially reduce their greenhouse gas (GHG) emissions. In the first period determined in the protocol, Germany committed itself to a reduction of 20% in GHG emissions between 2005 and 2012, which was already surpassed by 5% in 2010 (cf. Federal Ministry for the Environment, Nature Conservation, Building and Nuclear Safety, 2015). Nonetheless, in the second period a further reduction by

20% until 2020 has been agreed upon, which may require even stricter regulations. As mentioned before the macro-environmental changes that produce the requirement for an adaptation of current mobility systems are highly interrelated and strongly affect each other. Since urbanisation, and thus traffic, in urban areas increases municipalities are forced to proactively reduce traffic levels. Motorised private traffic has to be reduced, in particular, to tackle both environmental pollution as well as congestion and its negative health effects.

As price is one of the strongest levers to influence consumer behaviour, it could be argued that measures to increase the cost of MPT will be introduced in order to create an impetus for citizens to shift to other transport modes. One measure to achieve this may be the introduction of low emission zones which have been implemented in many German cities. This may increase the cost of MPT indirectly due to the gradual increase of emission norms and hence the prohibition of older vehicles. This could force people to purchase newer and more efficient models. Another example may be so-called congestion charges that are already implemented in other European cities such as London. Black (2010, p. 119) states that after the introduction of such a charge in London a reduction in traffic by around 20% was observed. A third measure maybe the increase in taxation of petrol which allows an almost instant increase in the cost of MPT.

Besides those punitive measures, the German government decided upon incentives for switching to more environmentally friendly vehicle engine types. The target of Germany is a fleet of 1,000,000 electric cars by 2020. To create an impetus for reaching this challenging goal owners of electric cars presently benefit from temporal tax exemptions (cf. Federal Government, 2016). The allowance of electric cars on dedicated bus lines as well as a reduction in fees for parking spaces is in discussion (cf. Federal Government, 2015).

Deffner and Götz (2010, p. 16) raise the importance of a holistic sustainable development approach. The authors argue that besides the protection of the environment and the reduction of GHG future mobility systems have to reduce mobility impairments and foster the feasibility to pursue socioculturally diverse lifestyles including their entailed mobility patterns.

In addition, several studies observed a general change in consumer behaviour. Capgemini (2013, p. 5) claim that besides the government, companies as well as individuals increasingly take the effects of their decisions on the environment into account. Krapf et al. (2013, p. 4) and Witzke (2016, p. 26) mention similar observations regarding the growing sensitivity towards the relationship between consumption patterns and the impact on the environment. Witzke (2016, p. 26) specifically refers to the consideration of an individual's mobility behaviour and its influence on the environment which is increasingly observed. Ahrens et al. (2011), as well as Rammler and Sauter-Servaes (2013), claim that there might be a societal shift towards more sustainable lifestyles. Ahrens et al. (2011, p. 22) state that values such as sustainability and rationality are increasingly important in the consumers' decision-making. Rammler and Sauter-Servaes (2013, p. 25) mention the occurrence of new lifestyle philosophies and consumption paradigms such as the lifestyle of health and sustainability (LOHAS), which combines environmental consciousness and the trend of physical fitness. Seitz (2013, p. 14) estimates the share of LOHAS in the German population to around 25 mn, which equals 37% of the total population from 14 years onwards. Moreover, this trend might not be limited to Germany. Transportation for America (2010, p. 11) observed similar developments in the United States.

Notwithstanding, Hopkins and Stephenson (2015, p. 13) found out that environmental motivations might not be the sole motivation to reduce MPT or rather behave environmentally conscious. Accordingly, additional conditions such as personal cost constraints still have a large influence on decisions such as transport mode choice.

2.4   Components of Digital Mobility

Similar to the discussion on the exact terminology of digital mobility, the concept's scope has also varied among different authors. For the purpose of this research previous publications have been reviewed in order to identify the most prominent components which will be outlined in the following paragraphs. It may be stated that although the digitalisation of mobility is rewarded the attribute to cause disruptive changes in the industry, the major elements of current systems will prevail. Digital mobility should rather be seen as a novel approach in the usage of these elements as well as their stronger integration enabled by the progressive implementation of new technologies and the digitisation of services (e.g. automation of metro systems or digital ticketing). Therefore, it was waived to include traditional components of mobility systems as well as their development in this chapter in order to avoid an unnecessary inflation of this research.

2.4.1   Status Quo Digital Mobility Service Components

Since mobility is finding itself in a transitional stage certain digital mobility components are already found in current transport systems. However, those components have not been acknowledged as integrative parts of the system by all consumers. This might be traced back to the limited integration of these components for example in terms of easiness in accessibility which may change in course of the industry's transformation. The majority of those elements are various types of vehicle sharing services. Vehicle sharing services are a vital element of digital mobility to ensure the same degree of flexibility car owners is savouring by simultaneously reducing the total amount of vehicles deployed.

Several studies have proven that car sharing users are as mobile and flexible as car owners when sufficient service coverage is given (cf. Wolter, 2012, p.540). Notwithstanding, at present state the effective usage of car sharing services is often hindered by such a lack of network density, flexibility (station-based car sharing), and administrative hurdles (i.e. registration and registration fees). The awareness of car sharing as an alternative to private cars has negative impact on the service usage as well (cf. Witzke, 2016, p. 14). Nonetheless, service operators are responding to the customers' requirements and progressively expanding their service coverage as well as simplifying the transaction process. Moreover, free float car sharing is increasingly introduced which enables users to pick-up and drop-off rented vehicles anywhere in the service region rather than at specific stations.

Another major sharing service type is ridesharing. Ridesharing, in contrast to car sharing, is mainly conducted on a peer-to-peer basis. The pairing process is usually carried out through Internet platforms such as BlaBlaCar. Users who offer seat capacities publish their trip onto the platform. Subsequently, ride seekers send requests to reserve a seat as well as to agree upon pick-up and drop-off locations. The payment is received in cash during the trip or paid online beforehand in the reservation process and transferred to the driver after the completion of the trip (cf. BlaBlaCar, 2016). Transportation for America (2010, p. 31) raises the importance of so-called dynamic ridesharing which has been pioneered in Ireland.[3] With this concept drivers and ride-seekers could be paired in real time without the necessity to prearrange pick-up and drop-off locations. Drivers would merely publish their route on the network and the system would automatically pair seekers en route with the same or nearby destination.

Furthermore, sharing services for non-motorised private transport have been introduced in many urban transport systems. Similar to car sharing, bike sharing is currently offered in two degrees of flexibility: station-based and free floating. Germany's largest bike sharing service operator Call a Bike was launched in the year 2000 in Munich as a phone-based bicycle rental system (cf. Witschel & Souren, 2014, p. 4). After its acquisition by German Railways Call a Bike expanded into 16 German cities offering both free-floating and station-based services. The rental process is still initiated by calling the operator's hotline. The customer is then requested to communicate the bike's identification number and receives a temporary pin code to unlock the bicycle. The drop-off is confirmed by simply pressing a button on the bicycle's lock. Depending on the bike sharing type offered, bicycles may be dropped-off anywhere in the service region in lieu of a service charge (cf. DB AG, 2010). Nevertheless, the registration via telephone or smartphone application is required at present state (cf. DB AG, 2010) which may reduce the attractiveness of those services.

In order to provide a complete overview of the status quo digital mobility components on-demand ride services (ODRS) should be mentioned as well. Companies such as Uber and Lyft provide platforms for services where customers can arrange car pick-ups via smartphone applications. In contrast to ridesharing services, the price paid to the driver exceeds the costs caused by the user. Thus, ODRS are comparable to regular taxi services. A major difference, however, maybe that most drivers are private persons. Nevertheless, due to Germany's strongly regulated taxi market Uber was forced by court decision to cease operation in Germany (cf. Handelsblatt, 2015).

---

[3]The Avego Project was initiated by the University College Cork to effectively broaden the public transit network by the integration of private cars.

2.4.2   Individualised Pricing Options

If one recalls the currently available pricing options of public transport or shared service operators it could be stated that an individual on average is able to choose between four options:

1. Trip-based pricing is based on fixed prices depending on the distance between the point of departure and destination or crossed fare zones.
2. Flat rate pricing is valid for an unlimited number of trips within an operator's network or parts of the network within a certain time period (e.g. day, week, month and year).
3. Trip-based and time-bound pricing such as tickets only valid in off-peak hours (e.g. 9 o'clock ticket of the Rhein-Main-Verkehrsverbund).
4. Time-bound flat-rate pricing which shares the characteristics of number three, however, enables the possessor to perform an unlimited number of trips for a certain time period (e.g. the Rhein-Main-Verkehrsverbund 9 o'clock monthly season ticket).

Although it might be stated that an individual is able to select one of the above-mentioned pricing schemes according to its needs and thus can personalise its pricing option to a certain extent, a true individualisation is unfeasible. For instance, with the purchase of a monthly season ticket a person automatically obtains the right of unlimited usage of all transport means provided by the respective operator in the respective zone. Notwithstanding, there might not be sufficient coverage of trams in that region or the possessor of the ticket does not intend to utilise such transport means. Thus, the person might pay for services he or she does not expect to use. This may increase the ticket price in the view of the customer and consequently reduce the attractiveness of the service offering. Furthermore, these flat rates are usually valid for the transport system of one operator only, making a 'true' flat rate unfeasible. An integrative part of digital mobility, therefore, is the personalisation of pricing options for both flat rates as well as trip-based tickets.

In the literature, two approaches similar to currently available pricing schemes are discussed. Specifically advocated by the Finnish Transport Agency (2015) and Heikkilä (2014) are tailor-made mobility packages according to the users' requirements. As pointed out by Hietanen (2014, 2 f.) users may be able to select between several packages which solely include transport and related services that are of interest to the customer. Kamargianni et al. (2015, p. 56) specify the package design by arguing that the customisation might be based on elements such as individual mobility patterns (e.g. current mode usage), socio-economic status (e.g. age, gender and family status) as well as attitudes and perceptions (e.g. lifestyle or environmental concerns). Moreover, there may be different types of service level agreements (SLA) between the customer and mobility operators which guarantee certain pick-up times after an order placement. Nonetheless, the 'Helsinki Model', as it is described by Kamargianni et al. (2015, p. 19), is not limited to such flat rates. It might also allow customers to make use of 'pay as you go' options (PAYGO).

PAYGO may be seen as the equivalent of the currently available trip-based tickets. A major difference however might be the ability to design door-to-door trips by using different transport modes of different operators. There are two types of those options currently discussed: dynamically and statically priced PAYGO. Fishman (2012, p. 15) and Goodall et al. (2015, p. 11) claim that the price per trip paid by the user might be dependent on several variables such as time of the day, congestion levels, speed, and occupancy. On the other hand, prices might be static and users will only pay according to their actual usage (e.g. km of shared car usage and distance travelled with public transport). It should be stated that the door-to-door design of mobility chains is not limited to the usage of public transport. The private car may be an integral part of the overall chain as well.

2.4.3 Autonomous On-Demand Vehicles

The introduction of autonomous driving vehicles (ADV) in digital mobility systems could contribute to the solution of various challenges especially urban mobility systems are currently facing. First of all, through the deployment of ADV it is expected that the amount of car-related incidents may reduce significantly (cf. International Transport Forum, 2015, p. 10; Goodall et al., 2015, p. 16). Thus, an improvement in road safety could be achieved. Fully automated vehicles might also aid in increasing the efficiency of the overall transport systems by avoiding congestion and improving the traffic flow. Moreover, such vehicles might drastically reduce the costs per trip. Currently, cars in Germany are used on average by 1.5 persons with an average usage time in most cases below 1 h per day (cf. Rammler & Sauter-Servaes, 2013, p. 33). Despite their massive and inefficient underutilisation many households still value private car ownership due to perceived benefits in terms of comfortable and schedule-less door-to-door travel (cf. International Transport Forum, 2015, p. 9). Autonomous mobility on-demand ride services (AMODS) may be a more cost-effective alternative.

AMODS might be seen as the successor of conventional car sharing services. Lenz and Fraedrich (2015) discuss two major types of AMODS in future mobility systems. The first type is referred to as valet-parking services. In this service-type vehicles take over the control from their initial location to the requesting user as well as during the search for parking space and en route to the next customer. During the actual utilisation the control is handed over to the user (cf. Lenz & Fraedrich, 2015, p. 185). The full vehicle-on-demand service on the other hand is based on level 4 automation. The request and the drop-off procedure are similar to valet-parking services. Nonetheless, during the utilisation the vehicle does not hand over the control to the user. Lenz and Fraedrich (2015, p. 187) compare this type of service to train compartments on the road, since the user is able to use the travelling time according to his or her preference such as reading, working or consumption of entertainment services. Mitchell et al. (2010, p. 49) on the contrary argue that

ADV may let a user switch to manual driving mode at any time. Since AMODS vehicles are shared an increase in utilisation can be achieved. This may result in lower costs per kilometre and thus will lead to more affordable transport solutions. In addition, costs for maintenance and other running costs associated with car owner-ship may become obsolete as well.

Furthermore, ADV could contribute to an increase in service level of public transport operators. The so-called first and last mile (i.e. the distance of a trip public transport users have to cover before a station of the respective transport mean is reached and vice versa) has been identified as one of the major factors for a negative perception of public transport service (cf. e.g. Siemens AG, 2015, p. 24; Tan & Tham, 2014, p. 6). Therefore, ADV may be used for these parts of the trip. Moreover, pick-up times could be synchronised with public transport departures to enable just-in-time drop-offs even when the subsequent transport mode is suffering from delays. Pavone (2015, pp. 408–410) simulated the fleet requirements of such a scenario for the cities of New York and Singapore. He concludes that for New York during peak hours the pick-up time could be reduced to 2.5 min with a fleet size of 8000 vehicles. For Singapore 300,000 vehicles could be sufficient during peak times to realise pick-ups after 15 minutes, which equals a reduction of deployed vehicles of around 61%. Fishman (2012, p. 24) even states that ADVs operated in platoons on dedicated tracks could match or exceed passenger throughput of public transport systems. Hence, ADV could also establish itself as a new mode of door-to-door public transport.

It might also be claimed that ADV will substitute regular private cars in the long term. Nevertheless, the cultural barrier towards ADV should not be neglected. Kröger (2015, p. 64) argues that the cultural shift from an automobile centred society towards a 'being chauffeured culture' itself might be the most significant challenge. Accordingly, this may be due to the fact that the automation of driving does not solely represent the alleviation of physical intense labour as in other automation cases. The author claims that with the automation of cars the joy of driving cars some individuals experience erodes as well. Therefore, it may be claimed that privately owned ADV will mostly provide valet services or merely take over control in unpleasant situations such as traffic jams or sections of long-distance trips.

2.4.4 Value-Added Services

Although various publications have acknowledged non-mobility services as element of digital mobility (cf. e.g. Capgemini, 2013, 8 f.; Cuddihey et al., 2015, p. 8), the actual services were not discussed in greater detail. This might be due to the variety of services that qualify for the integration into digital mobility systems, traced back to the strong focus on customer centricity and individuality in service provision. Therefore, it was opted to complement currently outlined services with potential value-added services (VAS) derived from the societal changes outlined in Sect. 2.3 that are seen as key drivers for digital mobility.[4]

The first feasible category may be entertainment services. The usage of smartphones and other mobile devices for entertainment purposes has increased rapidly over the last decade. A research conducted by Axel Springer Media Impact (2013, pp. 12–14) found out that 'infotainment' applications (e.g. news and information or television programme streams) are not only used daily by 51% of the participants under 35 years, but also account for one of the longest usage times. Multimedia applications in general were used by 54% on a daily basis in this quartile. It could be claimed that by now the usage of entertainment services can be frequently observed on public transport modes. Nevertheless, the evolvement of smart cars or rather ADV might also create a larger demand for the usage of those services in private or shared cars. At current state vehicles already possess interfaces for entertainment applications, Internet access or are equipped with DVD players (mostly for co-drivers). The usage of those services however may be seen as limited due to legal restrictions for the utilisation of entertainment equipment during vehicle operation, for example. As smart cars drive autonomously the travelling time might be used for entertainment or work purposes, thus creating demand for such services. According to Wehinger and Cords (2015, p. 145) 80% of new car buyers are already interested in complementing services. Cornet et al. (2012, p. 12) estimate development of the market for mobile media services and advertisement in smart cars to be 5 billion euros by 2030. Furthermore, by leveraging the potential of big data analytics, entertainment services could be offered proactively according to the customers' preferences and in coordination with the remaining trip length—even across multimodal mobility chains.

A second potential category may be seen in location-based commercial services. The principles of next best offer (NBO) marketing may become vital for appropriately offering customer-specific services. NBO is a personalised proposal of action which is founded on a customer's attributes, behaviour, purchase context (e.g. brick and mortar store or online) and product or service characteristics (cf. Davenport et al., 2011, 85 f.). Through the application of NBO marketing special offers at the user's original location and destination could be integrated into the mobility chain design. This could include offers at a customer's favourite coffee shop, but also seasonal sales at a branch of a preferred clothing company. The inclusion of tourist attractions or other places of interest to the customer is discussed as well (cf. e.g. Cuddihey et al., 2015, p. 8). Austin and Zielinski (n.d., p. 15) even claim that the provision of information on shops and services en route as well as on travel times and delays increases the attractiveness of public transport means. Furthermore, since a stronger integration of logistics and mobility services is aimed for (cf. e.g. Kunze, 2015, p. 8 f.), on-trip online shopping activities and the resulting parcel deliveries could be synchronised. One opportunity to achieve this might be seen in car boot deliveries. Audi in cooperation with DHL has been testing such

---

[4]Please note that the following paragraphs do not claim to be exhaustive.

delivery options (cf. Williams, 2015). Volvo has already launched similar services in Sweden (cf. Harris, 2015). This could enable customers to receive their parcels even while travelling. The customer's smart car could agree with the delivery service provider upon a certain place where the parcel should be handed over. Since same day delivery services are increasing in area coverage (see e.g. Amazon, 2016), the reception of a delivery at the trip's destination may be feasible. In addition, this may not solely apply to consumer goods, but also to groceries and fresh foods, where home deliveries are already available in many areas in Germany.

The enclosure of universal design (UD) principles into digital mobility services may be reasonable as well. Universal design is understood as *'the design of transport systems in a way that they are accessible to all users, irrespective of the users' abilities'* (Odeck et al., 2010, p. 304). UD might be utilised to balance a user's impairments such as physical constraints or lack of language proficiency. Users could be navigated through transport systems according to their needs (e.g. train exits equipped for disabled) or in their native tongue when travelling in a foreign country.

References

Ahrend, C., Schwedes, O., Daubitz, S., Böhme, U., & Herget, M. (2013). *Kleiner Begriffskanon der Mobilitätsforschung* (IVP-Discussion Paper No. 2013 (1)). Technische Universität Berlin.

Ahrens, G.-A., Bäker Berndard, Fricke, H., Körfgen, R., Schlag, B., Stephan, A., Wieland, B. (2011). *Zukunft von Mobilität und Verkehr: Auswertung wissenschaftlicher Grunddaten, Erwartungen und abgeleiteter Perspektiven des Verkehrswesens in Deutschland* (Forschungsbericht FE-Nr.: 96.0957/2010/). Technical University of Dresden.

Amazon. (2016). *Heute bestell. Heute da.* Retrieved August 10, 2021, from https://www.amazon.de/b?node=7013565031

Ammoser, H., & Hoppe, M. (2006). *GLOSSAR VERKEHRSWESEN UND VERKEHRSWISSENSCHAFTEN* (Diskussionsbeiträge aus dem Institut für Wirtschaft und Verkehr No. 2/2006). Technical University of Dresden.

Andrews, B. R. (1903). Habit. *The American Journal of Psychology, 14*(2), 121–149.

ARD/ZDF. (2016). *ARD/ZDF-Onlinestudie.* Retrieved May 06, 2021, from http://www.ard-zdf-onlinestudie.de/index.php?id=541

ARL. (2005). *Handwörterbuch der Raumordnung* (4th ed.). VSB Verlagsservice Braunschweig GmbH.

Austin, J., & Zielinski, S. (n.d.). *Building the new mobility economy: And supplying the emerging new mobility solutions for an urbanizing world.* University of Michigan.

Automotive Council UK. (2011). *Intelligent mobility: A National Need?* Automotive Council UK.

Axel Springer Media Impact. (2013). *Mobile Impact Academy I: Smartphone-Nutzung in Deutschland.* Axel Springer.

Bartz, F. M. (2015). *Mobilitätsbedürfnisse und ihre Satisfaktoren. Die Analyse von Mobilitätstypen im Rahmen eines internationalen Segmentierungsmodells.* (Doctoral Thesis). University of Cologne.

Becker, U. (2011). Verkehr und Umwelt: Zu den übergeordneten Zielen von Verkehrspolitik und der Rolle von Umweltaspekten. In O. Schwedes (Ed.), *Verkehrspolitik* (pp. 77–89). VS Verlag für Sozialwissenschaften.

BlaBlaCar. (2016). *So funktioniert's.* Retrieved May 27, 2021, from https://www.blablacar.de/wie-es-funktioniert

Black, W. R. (2010). *Sustainable transportation - problems and solutions* (1st ed.). The Guildford Press.

Bratzel, S. (2014). Die junge Generation und das Automobil – Neue Kundenanforderungen an das Auto der Zukunft? In B. Ebel & M. Hofer (Eds.), *Automotive Management. Strategie und Marketing in der Automobilwirtschaft* (2nd ed., pp. 93–108). Springer.

Buscher, V., Doody, L., Webb, M., & Aoun, C. (2014). *Urban-Mobility: Urban mobility in the smart city age (SMART cities cornerstone series)*. Arup & The Climate Group.

Cannon, S., & Summers, L. H. (2014). *How Uber and the sharing economy can win over regulators*. Retrieved October 13, 2021, from https://hbr.org/2014/10/how-uber-and-the-sharing-economy-can-win-over-regulators/.

Capgemini. (2013). *Mobilität der Zukunft: Reisen wird zum entspannten Erlebnis durch intermodale, integrierte und digitale Lösungen*. Capgemini.

Cerwenka, P. (1999). Mobilität und Verkehr: Duett und Duell von Begriffen? *Der Nahverkehr., 5,* 34–37.

Ciftci, K., Michel, A., & Siegfried, P. (2022). *The potential Impact of E-mobility on the Automotive value chain*. Springer Briefs in Business.

Cohen, B., & Kietzmann, J. (2014). Ride on!: Mobility business models for the sharing economy. *Organization & Environment, 27*(3), 279–296. https://doi.org/10.1177/1086026614546199

Cornet, A., Mohr, D., Weig, F., Zerlin, B., & Hein, A.-P. (2012). *Mobility of the future: Opportunities for automotive OEMs (Advanced industries)*. McKinsey & Company.

Corwin, S., Vitale, J., Kelly, E., & Cathles, E. (2015). *The future of mobility: How transportation technology and social trends are creating a new business ecosystem*. Deloitte University Press.

Cuddihey, A., Butler, S., Schneider, E., Sim, H., & Wilson, M. (2015). *The AccentureTechnology vision 2015: Public transportation: Riding the digital era*. Accenture.

Cullinane, S., & Edwards, J. (2010). Assessing the environmental impacts of freight transport. In A. McKinnon, S. Cullinane, M. Browne, & A. Whiteing (Eds.), *Green logistics - improving sustainability of logistics* (pp. 31–48). Kogan Page.

Davenport, T. H., Mule, D. L., & Lucker, J. (2011). Know what your customers want before they do. *Harvard Business Review, 89*(12), 84–92.

DB AG. (2010). *Call a Bike in Frankfurt*. Retrieved May 28, 2021, from https://www.callabike-interaktiv.de/index.php?id=396&&f=500

Deffner, J., & Götz, K. (2010). *DIE ZUKUNFT DER MOBILITÄT IN DER EU*. European Parliament.

Delatte, A., Kettner, S., Schenk, E., & Schuppan, J. (2014). *Multimodale Mobilität ohne eigenes Auto im urbanen Raum: Eine qualitative Studie in Berlin Prenzlauer Berg*. Technical University of Berlin.

DuPuis, N., Martin, C., & Rainwater, B. (2015). *City of the future: Technology & mobility*. National League of Cities.

Ebel, B., Hofer, M., & Genster, B. (2014). Automotive management – Trends und Ausblick für die Automobilindustrie. In B. Ebel & M. Hofer (Eds.), *Automotive management. Strategie und Marketing in der Automobilwirtschaft* (2nd ed., pp. 539–548). Springer.

Elkington, J. (1997). *Cannibals with forks - The triple bottom line of 21st century business* (1st ed.). Capstone Publishing Limited.

EMI. (2012). *A strategic knowledge and research agenda on sustainable urban mobility*. European Metropolitan Network Institute.

Eriksson, L. (2011). *Car users' switching to public transport for the work commute*. Karlstad University.

Federal Government. (2015). *Merkel: Elektroautos weiterhin fördern*. Retrieved February 06, 2021, from https://www.bundesregierung.de/Content/DE/Artikel/2015/06/2015-06-15-bkin-elektro-konferenz.html

Federal Government. (2016). *Leitmarkt und Leitanbieter für Elektromobilität*.

Federal Ministry for the Environment, Nature Conservation, Building and Nuclear Safety. (2015). *Kyoto-Protokoll.* Retrieved May 28, 2021, from http://www.bmub.bund.de/themen/klima-energie/klimaschutz/internationale-klimapolitik/kyoto-protokoll/

Federal Statisical Office. (2009). *Germany's population BY 2060: Results of the 12th coordinated population projection.* Federal Statistical Office of Germany.

Finnish Transport Agency. (2015). *MaaS services and business opportunities* (Research reports of the Finnish Transport Agency No. 56/2015). Finnish Transport Agency.

Firnkorn, J. (2012). Triangulation of two methods measuring the impacts of a free-floating carsharing system in Germany. *Transportation Research Part A: Policy and Practice, 46*(10), 1654–1672. https://doi.org/10.1016/j.tra.2012.08.003

Fishman, T. D. (2012). *Digital-age transportation: The future of urban mobility.* Deloitte University Press.

Goodall, W., Fishman, T., Dixon, S., & Perricos, C. (2015). *Transport in the digital age: Disruptive trends for smart mobility.* Deloitte.

Goudin, P. (2016). *The cost of non-Europe in the sharing economy: Economic, social and legal challenges and opportunities* (No. PE 558.777). European Parliamentary Research Service (EPRS).

Groß, S., & Freyer, W. (2010). *Mobilitäts- und Verkehrsverhalten von Jugendlichen.* Technische Universität Dresden.

Grotenhuis, J.-W., Wiegmans, B. W., & Rietveld, P. (2007). The desired quality of integrated multimodal travel information in public transport: Customer needs for time and effort savings. *Transport Policy, 14*(1), 27–38. https://doi.org/10.1016/j.tranpol.2006.07.001

Handelsblatt. (2015). *Uber in ganz Deutschland verboten.* Retrieved January 07, 2021, from http://www.handelsblatt.com/unternehmen/dienstleister/umstrittener-fahrdienst-uber-in-ganz-deutschland-verboten/11522380.html

Harris, J. (2015). Volvo starts delivering goods right to your car's boot – even if you're not there. Retrieved January 06, 2021, from http://home.bt.com/tech-gadgets/tech-news/volvo-deliver-online-orders-car-boot-11364020943305

Heikkilä, S. (2014). *Mobility as a service – A proposal for action for the public administration. Case Helsinki* (Master's thesis). Aalto University.

Henkel, S., Tomczak, T., Henkel, S., & Hauner, C. (2015). *Mobilität aus Kundensicht.* Springer Fachmedien Wiesbaden.

Hietanen, S. (2014). 'Mobility as a service' – The new transport model? *Eurotransport, 12*(2), 2–4.

Hopkins, D., & Stephenson, J. (2015). *Generation Y mobilities: Highlights.* University of Otago.

Huber, A., & Laverentz, K. (2012). *Logistik.* Vahlen Verlag.

imfo. (2011). *Mobilität junger Menschen im Wandel – multimodaler und weiblicher* (ifmo Studien). Institut für Mobilitätsforschung (imfo).

International Transport Forum. (2015). *Urban mobility system upgrade: How shared self-driving cars could the city traffic.* International Transport Forum.

Kamargianni, M., Matyas, M., Li, W., & Schäfer, A. (2015). *Feasibility study for "mobility as a service" concept in London: FS-MaaS project – Final deliverable.* UCL Energy Institute.

Krapf, H., Fischer, K., Wehlau, D., & Weller, I. (2013). *Klimawandel im Alltag: Neue Impulse für nachhaltigen Konsum?* (rtec-paper No. 193). University of Bremen.

Kröger, F. (2015). Das automatisierte Fahren im gesellschaftsgeschichtlichen und kulturwissenschaftlichen Kontext. In M. Maurer (Ed.), *Autonomes Fahren. Technische, rechtliche und gesellschaftliche Aspekte* (pp. 41–67). Springer.

Kuhnimhof, T., Buehler, R., Wirtz, M., & Kalinowska, D. (2012a). Travel trends among young adults in Germany: Increasing multimodality and declining car use for men. *Journal of Transport Geography, 24*, 443–450. https://doi.org/10.1016/j.jtrangeo.2012.04.018

Kuhnimhof, T., Wirtz, M., & Manz, W. (2012b). Decomposing young Germans' altered car use patterns. *Transport Res Rec, 2320*, 64–71. https://doi.org/10.3141/2320-08

Kunze, W. M. (2015). Connected car und smart connected delivery Potenziale & Innovationen: Car sharing, autonome autos, smarte Lieferservices und neue Geschäftspotenziale für automotive, Handel, big data und startup-Modelle. trendquest.

Lauwers, D., & Papa, E. (2015). *Towards a smarter urban mobility* (Contribution to colloquium Vervoersplanologisch Speurwerk). Antwerp

Lempp, M., & Siegfried, P. (2021). *Automotive disruption and the urban mobility revolution - Rethinking the business model 2030, business guides on the go*. Springer.

Lenz, B. (2011). Verkehrsrelevante Wechselwirkungen zwischen Mobilitätsverhalten und Nutzung von IuK-Technologien. *Informationen zur Raumentwicklung, 10*, 609–618.

Lenz, B., & Fraedrich, E. (2015). Neue Mobilitätskonzepte und autonomes Fahren: Potenziale der Veränderung. In M. Maurer (Ed.), *Autonomes Fahren. Technische, rechtliche und gesellschaftliche Aspekte* (pp. 174–195). Springer.

Metz, D. (2013). Peak car and beyond: The fourth era of travel. *Transport Reviews, 33*(3), 255–270. https://doi.org/10.1080/01441647.2013.800615

Mitchell, W. J., Borroni-Bird, C., & Burns, L. D. (2010). *Reinventing the automobile: Personal urban mobility for the 21st century*. Massachusetts Institute of Technology.

Musti, S., & Kockelman, K. M. (2011). Evolution of the household vehicle fleet: Anticipating fleet composition, PHEV adoption and GHG emissions in Austin, Texas. *Transportation Research Part A: Policy and Practice, 45*(8), 707–720. https://doi.org/10.1016/j.tra.2011.04.011

National IT Summit. (2015). *Kompass Digitale Netze und intelligente Mobilität: Poten ziale erkennen, Richtung bestimmen*. National IT Summit.

Odeck, J., Hagen, T., & Fearnley, N. (2010). Economic appraisal of universal design in transport: Experiences from Norway. *Research in Transportation Economics, 29*(1), 304–311. https://doi.org/10.1016/j.retrec.2010.07.038

Otto, S. (2010). *The psychology of transport choice*. Institute for Ecological Economic Research (IÖW).

Pavone, M. (2015). Autonomous mobility-on-demand systems for future urban mobility. In M. Maurer (Ed.), *Autonomes Fahren. Technische, rechtliche und gesellschaftliche Aspekte* (pp. 399–416). Springer.

PwC. (2015). *Share Economy: Repräsentative Bevölkerungsumfrage*. PricewaterhouseCoopers (PwC).

Rammler, S., & Sauter-Servaes, T. (2013). *Innovative Mobilitätsdienstleistungen* (Arbeitspapier No. 274). Hans-Böckler-Stiftung.

Scheiner, J. (2007). Mobility biographies: Elements of a biographical theory of travel demand. *Erdkunde, 61*(2), 161–173.

Seitz, J. (2013). *FAIR - Von der Nische zum Mainstream*. Zukunftsinstitut.

Siegfried, P. (2021). *Land & sea transport - aviation management*. BoD Book on Demand.

Siemens AG. (2015). [Special issue]. *como - Fakten, Trends und Stories zu integrierter Mobilität*. (14). Gutenber Beuys.

Streit, T., Chlond, B., Weiß, C., & Vortisch, P. (2015). *Deutsches Mobilitätspanel (MOP) – Wissenschaftliche Begleitung und Auswertungen Bericht 2013/2014: Alltagsmobilität und Fahrleistung* (Forschungsprojekt FE-Nr. 70.0864/2011)70.0864/2011). Karlsruher Institut of Technology (KIT).

Tan, C. K., & Tham, K. S. (2014). Autonomous vehicles, next stop: Singapore. *Journeys, 12*, 5–10.

TNS Emnid. (2015). *Sharing economy: Die Sicht der Verbraucherinnen und Verbraucher in Deutschland*. Verbraucherzentrale Bundesverband (vzbv).

Transportation for America. (2010). *Smart mobility for a 21st Century America: Strategies for maximizing technology to minimize congestion, reduce emissions, and increase efficiency* (White Paper). Transportation for America, ITS America, Association for Commuter Transportation, & University of Michigan's SMART Initiative.

Tully, C. J., & Baier, D. (2006). *Mobiler Alltag: Mobilität zwischen Option und Zwang - Vom Zusammenspiel biographischer Motive und sozialer Vorgaben* (1st ed.). VS Verlag für Sozialwissenschaften.

Umweltbundesamt. (2015). *Marktbeobachtung Nachhaltiger Konsum: Entwicklung eines Instrumentes zur Langzeit-Erfassung von Marktanteilen, Trends* (Texte No. 02/2015). Umweltbundesamt.

UNDESA. (2015). *World Population Prospects: The 2015 Revision* (Volume II: Demographic Profiles (ST/ESA/SER.A/380)). United Nations, Department of Economic and Social Affairs, Population Division.

Vredin Johansson, M., Heldt, T., & Johansson, P. (2006). The effects of attitudes and personality traits on mode choice. *Transportation Research Part A: Policy and Practice, 40*(6), 507–525. https://doi.org/10.1016/j.tra.2005.09.001

Wehinger, J., & Cords, S. (2015). Transformation von Geschäftsmodellen in der Automobilindustrie am Beispiel von "Automatischem Fahren". In H. Proff (Ed.), *Entscheidungen beim Übergang in die Elektromobilität* (pp. 143–153). Springer Fachmedien Wiesbaden.

Williams, R. (2015). Amazon to deliver packages to your car boot. Retrieved January 06, 2021, from http://www.telegraph.co.uk/technology/amazon/11558457/Amazon-to-deliver-packages-to-your-car-boot.html

Windzio, M. (2013). Räumliche Mobilität. In S. Mau & M. Schöneck Nadine (Eds.), *Handwörterbuch zur Gesellschaft Deutschlands* (3rd ed., Volume 1 and 2, pp. 663–675). Springer Fachmedien.

Witschel, J., & Souren, R. (2014). *Kapazitätswirtschaftliche Analyse der Strukturelemente und Determinanten des Bikesharing* ([Online-Ausg.]). *Ilmenauer Schriften zur Betriebswirtschaftslehre: Vol. 2014, 2*. Ilmenau, Ilmenau: Verl. proWiWi; Univ.-Bibliothek. Retrieved from http://nbn-resolving.de/urn:nbn:de:gbv:ilm1-2014200069

Witzke, S. (2016). *Carsharing und die Gesellschaft von Morgen: Ein umweltbewusster Umgang mit Automobilität?* Springer Fachmedien.

Wolter, S. (2012). Smart Mobility- Intelligente Vernetzung der Verkehrsangebote in Großstädten. In H. Proff, J. Schönharting, D. Schramm, & J. Ziegler (Eds.), *Zukünftige Entwicklungen in der Mobilität. Betriebswirtschaftliche und technische Aspekte* (1st ed., pp. 527–548). Springer Fachmedien.

Xia, F., Yang, L. T., Wang, L., & Vinel, A. (2012). Internet of things. *International Journal of Communication Systems, 25*(9), 1101–1102. https://doi.org/10.1002/dac.2417

Zierer, M. H., & Zierer, K. (2010). *Zur Zukunft der Mobilität: Eine multiperspektivische Analyse des Verkehrs zu Beginn des 21.* Jahrhunderts (1. Aufl.). Springer Fachmedien.

Zobrist, L., & Grampp, M. (2015). *The sharing economy: Share and make money: How does Switzerland compare?* Deloi.

Digital Mobility Business Concepts

3

3.1 Changing Customer Requirements on Mobility Services

It has been anticipated by several studies that consumer behaviour will undergo massive changes until 2020 which may have disruptive impacts on various industries, including the mobility service providers'. Mostly fuelled by the enablers and drivers discussed in Sect. 2.3, companies are expected to face three major trends[1]: individualisation/customisation, commoditisation, and convenience.

It could be stated that individualisation is both driven by customers' desire for individuality and the result of personal constraints such as income or physical impairments. As noted by Winterhoff et al. (2009, p. 14) as well as Ternès et al. (2015, p. 13) individuals increasingly divert from classical lifestyle patterns, thereby creating demand for a variety of products which suit their specific lifestyle (e.g. LOHAS, best agers or ethic consumption). Specifically, younger generations are accustomed to choose between a large variety of goods and services (cf. Parment, 2013, p. 36) and thus may place high expectations on companies. Ternès et al. (2015, p. 27) claim that due to the rapid advance in technology—in particular, due to the currently ongoing digitisation—customisation of goods and services might gain even stronger importance. Bechmann et al. (2015, p. 5) for example mention the demand of younger adults for 'virtual customisation' of on-board computer and entertainment systems in cars through personalised services and user profiles.

Mass customisation may also have strong impact on commoditisation. As companies respond to the growing demand for individualisation, the variety of products increases while the diversity decreases. This homogenisation may lead to a vast amount of products and services which are solely distinguishable by the price they are offered at. Commoditisation is no new phenomenon and has already led many industries such as the transport or telecommunications service industry into

---

[1] Please note that the following list of trends is not exhaustive. Merely trends which demonstrated strong impact potential on mobility and transport services are discussed.

© The Author(s), under exclusive license to Springer Nature Switzerland AG 2022   33
P. Siegfried, *Digitalisation in Mobility Service Industry*, Future of Business and Finance, https://doi.org/10.1007/978-3-031-07151-5_3

price wars. Nonetheless, through the ubiquitous connectivity of consumers and the rapidly increasing information available online the retention of customers in highly commoditised markets may become even more challenging. Conroy et al. (2015, p. 4) found out that traditional measures to build customer loyalty are gradually losing their effectiveness and relevance. In their research, Accenture (2014, p. 3) observed similar developments and categorise the erosion of customer loyalty caused by digitalisation as the research most compelling findings. Accordingly, customers not only consider more options in their decision-making, but also steadily compare and evaluate the services of their current provider with those available on the market. Accenture (2014, 4f) further argue that although the participation in customer loyalty programmes is repeatedly observed, the customers' intention to participate might be short-termed and pure opportunistic to gain access to special offers, amongst others.

Digitalisation could also influence consumer behaviour in terms of convenience of service provision or fulfilment. Over the last decade online shopping has established itself as convenient alternative to brick-and-mortar stores due to ubiquitous and time-independent access to a large variety of products at often lower prices. A large-scale research by PwC (2016) of around 23,000 participants in 25 countries revealed that 54% shop online on a weekly or monthly basis. Moreover, 45% are already shopping via their mobile phones, which in future will become the main purchasing tool according to 34% of the participants. The research also found out that for the German market convenience is the main rationale for shopping online (54%) followed by price (35%). Nevertheless, the expectation of convenient service provision might impose several challenges on companies. Accenture (2014, p. 2) observed that customers have become more impatient and are less willing to put vast effort into the buying decision process. Subsequently, the change of providers due to a service provision which did not meet the customers' expectations in terms of, for instance, lack of compelling offers or a low 'digital intenseness', was commonly observed (cf. Accenture, 2014, pp. 3–5).

For mobility service providing companies in particular these three trends may have disruptive implications. It could be claimed that in terms of individualisation customers will demand service packages which only contain services they regard as useful. Moreover, the services must cover a large variety of options and have to be accessible easily and conveniently to reduce the effort required for the buying decision process. It could further be claimed that the services will have to be transport mode independent or transport mode comprehensive as the demand for multimodal mobility is continuously increasing (cf. Kuhnimhof et al., 2012, p. 444). In order to bind customers, companies may have to provide incentives that truly add value in the view of their clients. The impact of those trends, however, might be related to the degree service providers already face these challenges. It could be stated that companies such as airlines and car manufacturers have been exposed to the requirements to provide personalisation options as well as digitalise their service offerings for several years. Public transport operators on the other hand could be seen as laggard in these areas. Nevertheless, as supported by Conroy et al. (2015, p. 4) the manner business is conducted at current state has to undergo substantial adjustments to secure future economic viability, regardless of a company's current business success. The effort a company has to dedicate to the transformation may be dependent on the previously mentioned status quo, however.

3.2   Customer Centricity as Overarching Management Paradigm

The diverging and increasingly customer-specific requirements laid down on mobility providing companies combined with the strong commoditisation of mobility services might produce an even stronger necessity for customer-centric management (CCM). The principle of customer centricity, specifically represented in the customer relationship management (CRM) approach, has been introduced in the 1990s and rapidly advanced to the dominating marketing paradigm (cf. Winer, 2001, p. 89). Notwithstanding, due to the absence of a commonly accepted definition the degree of implementation has varied amongst companies, which led to a rather mock application of customer centricity principles. A research conducted by Allen et al. (2005) concluded that while 80% of the 362 surveyed companies stated that their business is aligned towards its customers merely 8% of their customers confirmed this allegation. More recent studies (cf. e.g. Accenture, 2016; HBR Analytical Services & Strativity, 2015) demonstrate that improvements on a larger scale have still not been achieved in this area. This could suggest that a mere application of CRM might not be sufficient to achieve genuine customer centricity.

In recent years, another concept to improve relationships between customers and companies has been developed and discussed among scholars and practitioners: The customer experience management (CEM) approach. At current state, the literature does not provide consensus concerning the relationship between CRM and CEM. Therefore, it seems crucial to briefly review the two concepts, their similarities, differences, and interrelations in order to identify the most appropriate alternative for the application in digitalised mobility businesses.

As mentioned before, CRM has experienced strongly varying interpretations. Chen and Popovich (2003, p. 673) state that the extent organisations define CRM ranges from a mere technology solution to a cross-functional, customer-driven, technology-integrated business process management strategy. Payne and Frow (2005, 167 f.) also underline these two extreme interpretations and see the incorrect equation of CRM and CRM technology solutions as a key failure of successful CRM applications. Nonetheless, the importance of technology within CRM is emphasised by the authors. Rosman and Stuhura (2013, p. 19) summarise CRM as a process that supports a company in profiling its (potential) customers in order to understand their specific needs and build relationships with them by offering the most suitable products or services. This emphasises the analytical nature of CRM. Malthouse et al. (2013, p. 270) state that through CRM a company intends to leverage gathered customer information to maximise customer lifetime value (i.e. the value of all future transactions between the customer and the firm). With every interaction a firm gains valuable information about its customer which is analysed and shared throughout the company to maximise customer profitability (cf. Kamaldevi, 2010, p. 38). Moreover, by actively managing customer relationships it is expected to reduce customer migration and thus costs for new customer recruitment or recovery.

CEM on the other hand aims on the emotional side of the customer–company relationship. According to Verhoef et al. (2009, p. 32), the customer experience comprises the total transaction lifecycle including the search, purchase, consumption and aftersales phase and requires the customer's involvement at different levels (rational, emotional, sensorial, physical and spiritual) due to its strictly personal nature. CEM hence intends to improve the total experience by gathering information on customer interactions to optimise every touch point and influence the customer behaviour at every opportunity (cf. Kamaldevi, 2010, p. 38). The optimisation of touch points should result in an emotional bond between the customer and the company or rather the services and products offered.

By reviewing the two CCM concepts several similarities can be identified which might also account for the failure to form a consensus on the exact relationship between CRM and CEM. Both concepts are technology-integrated solutions that aim to gather information about customer behaviour in order to predict feasible next actions and improve a business's profitability. Moreover, both concepts are utilised to personalise service offerings, and intend to bind a customer to the business by creating a long-term relationship. Nonetheless, views on the concepts' interrelation range from regarding CRM and CEM as two separate concepts sharing certain similarities, CEM being a more developed successor of CRM, CRM as a component of CEM, to CRM and CEM being an integrative part of a broader customer-centric management concept. Meyer and Schwager (2007, 4 f.) claim that CRM and CEM mostly differ in five areas: matter, timing, monitoring, audience and purpose. Accordingly, CRM is rather based on historic information which is gathered after an action (e.g. request of service or product return) was carried out to learn about the customer. CEM on the other hand immediately captures customers' subjective thoughts to add offerings in the gap between expectations and experience. These major differences between the two approaches are summarised in Table 3.1.

It should be stated that in the context of customer-centric digital mobility business models the pursuit of a holistic CCM approach (i.e. the usage of CRM and CEM) is crucial to fully satisfy the requirements of individualised service provision. This might have several reasons. Firstly, as pointed out by Kamaldevi (2010, p. 38) and Meyer and Schwager (2007, p. 5) CRM as well as CEM serve a specific purpose to deliver customer-centric services (cf. Siegfried, 2014, p. 85). While CRM supports businesses to profile their customers and learn which products are preferred, CEM enables companies to design the relationship from initiation to termination according to the customers' desires. This might highlight the dependence of both concepts on one another. It could be argued that without CRM data an optimal and individualised customer experience design is unfeasible. On the other hand, neglecting the customer experience and its impact on the overall relationship lifecycle may hinder companies to successfully retain its customers in strongly commoditised markets. Moreover, acknowledging user experience as a competitive advantage is especially

Table 3.1 Differences between CEM and CRM

	What	When	How monitored	Who uses the information
CEM	Captures and distributes what a customer thinks about a company	At points of customer interaction: 'Touch points'	Surveys, targeted studies, observational studies, 'voice of customer' research	Business or functional leaders, in order to create fulfillable expectations and better experiences with products and services
CRM	Captures and distributes what a company knows about a customer	After there is a record of a customer interaction	Point-of-sales data, market research, web site clickthrough, automated tracking of sales	Customer-facing groups such as sales, marketing, field service, and customer service, in order to drive more efficient and effective execution

Source: Own depiction

important in service-providing companies (cf. Berry et al., 2002, p. 1; Berry & Carbone, 2007, p. 26).

3.2.1   Status Quo Application

In order to gain a deeper insight into the current application of CRM and CEM the most prominent methods and techniques should be reviewed.

Generally, CRM is divided into three levels which ought to generate a 360-degree view on a company's customers by feeding and interacting with one another (cf. Payne, 2005, p. 23):

1. Operational CRM which is concerned with the automation of business processes such as marketing, sales and customer service.
2. Analytical CRM that captures, stores, analyses and interprets data gathered from the operational side.
3. Collaborative CRM which represents the infrastructure to make interactions between the customer and the company or its employees feasible throughout multiple channels.

Analytical CRM could be seen as the foundation of the overall construct. With the aid of online analytical processing (OLAP) as well as data mining techniques customers are categorised into specific segments to select the most appropriate operational CRM processes to approach the respective customer (cf. Leußer et al., 2011, p. 40). These segmentations are based on several variables which may include buying behaviour, demographics as well as anticipated monetary and non-monetary customer value. Moreover, segmentations are the basis for so-called cross-selling and upselling analyses. In the cross-selling analysis customers are offered complementing products which are founded on previous purchases or derived from association and sequence analyses that determine dependencies or purchase orders of certain products (cf. Leußer et al., 2011, p. 41). The upselling analysis identifies the potential to sell more profitable products or service upgrades to customers (Dyché, 2001, p. 20). These two analyses should not only aid to bind customers to the company, but also to transform less profitable customers into profitable ones (i.e. improving the customer lifetime value).

In the operational CRM, the insights gathered from analyses in the analytical component are utilised to select the most suitable approach to interact with the respective customers. This includes the selection of the most appropriate customer touch points and marketing channels as well as the choice of marketing campaigns which should generate interest at the customer side to trigger new purchases. Dyché (2001, 25 f.) mentions the usage of event-based marketing in this regard. Accordingly, firms should define a set of high-profile events which call for interactions with the customer to make marketing campaigns appear more personalised. Besides the personalisation of sales and marketing processes, customer loyalty programmes are frequently implemented to create incentives for new purchases as well as bind customers to the company. Loyalty programmes are often based on progressively gaining access to larger discounts or special treatments with every time a purchase is made.

It might be stated that CEM has a vast impact on a customer's perception of operational CRM processes. As an ultimate goal CEM aims to deliver a distinct customer experience by achieving service excellence. According to Homburg et al. (2013, p. 7) CEM follows two principles to achieve this. The touch point principle is based on influencing the customer at various contact points on several emotional and non-emotional levels. The experience principle on the other hand is concerned with designing and determining an order of events along the entire transaction lifecycle which costumers would regard as valuable, the so-called customer journey. Moreover, Homburg et al. (2015, pp. 13–15) stress the importance of three cornerstones CEM should be founded on: consistency, context sensitivity and connectivity of touch points. Accordingly, customers should perceive a consistency in terms of corporate identity at all touch points. In addition, touch points should be sensitive to the customers' situational context for example by offering child care or making desired dates for parcel deliveries available (cf. Homburg et al., 2015, p. 14). This may also highlight the necessity to integrate online and offline touch points to enable a seamless transition. As an example of such an integration Homburg et al. (2015, p. 14) mention the validity of vouchers for online and offline touch points as well as an overview of online and in-store stock availability.

Likewise CRM CEM heavily relies on data gathered from customer contacts. Nevertheless, a strong difference may be that CEM rather focuses on qualitative data, since customer experience is highly subjective. Meyer and Schwager (2007, 6 f.) state that companies may collect data at several opportunities or trigger points to ensure service excellence. The first opportunity may be a follow-up after a recent purchase by the customer to improve transactional experiences as well as assess the impact of new marketing initiatives. A second opportunity is seen in the collection of data on current relationship or rather experience issues by conducting 'follow them home' case studies or biannual account reviews. The third opportunity is to test future opportunities to improve the service offering with focus group interviews or special-purpose market studies.

3.2.2 Application and Challenges for Digital Age Mobility Providers

Digitalisation inherits the potential to improve the customer centricity of mobility service providers significantly. Through the increasing availability of customer-specific data companies might be able to segment its customers with enhanced accuracy. Thus, segmentation could be done on a genuine one-to-one basis rather than on demographics and buying behaviour. The personalisation of marketing campaigns as well as cross-selling and upselling analysis might also improve due to the availability of detailed data on customer preferences. In terms of mobility services, this may include the selection of transport means which are preferred by the customer as well as the avoidance of for example overcrowded train compartments. Besides, upselling to first-class compartments could be linked to vehicle occupancy for customers who prefer to travel more privately.

Furthermore, the service provision could be personalised as well. Currently, customer experience optimisation is based upon subjective impressions of a variety of customers. Nonetheless, by leveraging data gathered in CRM systems the service provision could be adapted to a customer's preferences. This could include the level of proactive service provision. Some travellers may prefer to rather receive information and services strictly concerned with transport services, while other customers might enjoy to proactively receive suggestions on VAS for example as described in Sect. 2.4.4.

The realisation of such a scenario however may face several challenges. One challenge can be seen in the generation of sufficient and qualitative data. If one recalls companies which demonstrate strong performances in CCM it could be claimed that those firms are operating in an environment, which can be referred to as a closed system. Airlines for example are in possession of data on their customers' location within the network at any time. Internet companies such as Google and technology firms such as Apple or Samsung can make use of location-based services and data on user preferences. For companies such as public transport operators, the obtainment of such data may be seen as rather challenging. Car manufacturers might own data on the performance and usage of their customers' cars. Nevertheless, gathering data on buying behaviour could be challenging as well. Therefore, those companies could face the challenge to transform their systems into a more closed one.

Another challenge could be seen in the willingness of various companies to share information about their customers. Similar to the management on supply chains, mobility chains may require information from all parties involved in the service provision not only to enable seamless travel, but also to provide VAS at all stages of the journey.

3.3   Business Model Archetypes for a Digitalised Mobility Market

In the respective literature, potential business model archetypes have been discussed which could permit the establishment of a holistic mobility provider (HMP). The HMP may act as an orchestrator of all related mobility and non-mobility services and is the main access point for designing customised door-to-door mobility chains in the digital mobility ecosystem. In general, three archetypes as well as their hybrid versions to serve as foundation for such a business are proposed.

The *integrational or bottom-up approach* relies upon the provision of services which are carried out by one or a limited amount of companies that act as a consortium of strategic partners. Companies that pursue this approach expand their current core business into new segments in order to be able to provide holistic mobility services. The expansion is achieved by heavily relying on deep (vertical) integration of additional services to assure seamless and fast multimodal travel (cf. van Audenhove et al., 2014, p. 22) which could be seen as the major advantage of the concept. Moreover, consistent quality in service provision might be feasible throughout the entire mobility chain. It could further be argued that the concept of servitisation is a crucial component of this approach as it is mostly applied by established companies that either produce or possess physical assets to provide transport services such as car manufacturers or railway companies (cf. Rammler & Sauter-Servaes, 2013, p. 40). The term servitisation was coined by Vandermerwe and Rada (1988, 315 f.) and describes the development of firms from merely focusing on providing products to the offering comprehensive bundles of products and complementing services. According to the authors, the shift towards selling the utilisation of a product rather than the product itself is increasingly gaining importance. Winterhoff et al. (2009, pp. 62–69) specifically discuss four archetypes for the servitisation in the automotive industry. They conclude that car manufacturers could position themselves as depicted in Fig. 3.1.

As demonstrated in the figure above companies may opt for the provision of services complementing their own product (Service Focused Manufacturer) or decouple the service provision completely from their own products (Mobility Service Provider). Except for the 'Product Focused Manufacturer', in all business models vehicles are leased or rented to the customers and complimentary services are offered based on the customers' individual necessities throughout the relationship lifecycle. The authors explicitly highlight the integration of non-mobility services in those archetypes (e.g. e-commerce services). The most radical transformation may be seen in the 'Mobility Service Provider' model, in which manufacturers completely focus on the management of their customers' mobility without selling any physical products. Nevertheless, due to the rather asset-driven background of integrators the creation and collection of customer-specific data on other aspects than the transport itself (i.e. to offer complementing services) could be seen as a major challenge for these businesses.

Furthermore, as services are provided by merely one or a few companies Lerner and van Audenhove (2012, p. 18) argue that besides a dense and extensive service

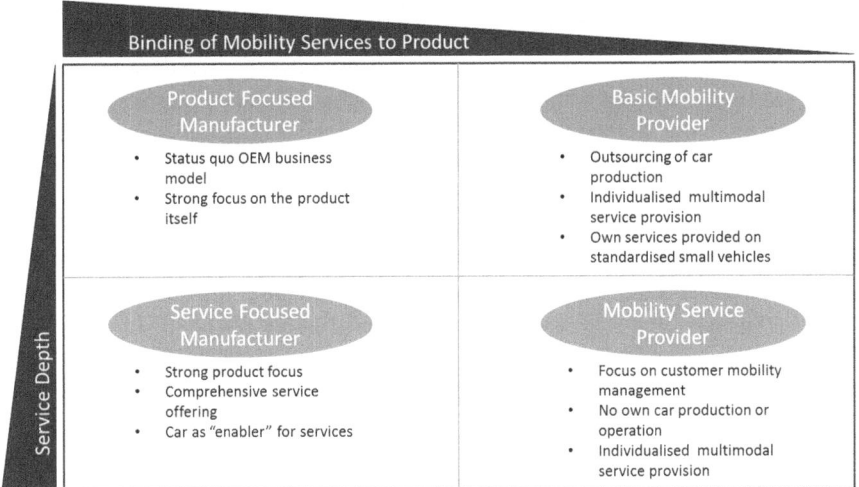

Fig. 3.1 Servitisation business models in the automotive industry (Source: Own depiction)

network a strong brand image may be crucial for the success of the integrational approach.

The *aggregational or top-down approach* rather focuses on providing a central point of access to the existing heterogeneous landscape of mobility services to increase the suitability for daily use (cf. Rammler & Sauter-Servaes, 2013, p. 45). Mostly by offering user-friendly interfaces for mobile devices such as smartphones aggregators enable their customers to design individual door-to-door mobility chains by accessing a variety of different service providers. Moreover, customers may be able to access information as well as compare routes, book and initiate payments (cf. van Audenhove et al., 2014, p. 22). In contrast to the integrational approach, aggregators do not require large investments into physical infrastructure or assets as the accumulation of third-party services is seen as a core competence. Therefore, aggregators may be strongly dependent on the cooperation of third parties. Moreover, consistent service quality levels might be hard to achieve due to the variety of entities involved in the service provision. Rammler and Sauter-Servaes (2013, p. 45) claim that this approach is mainly pursued by new entrants into the mobility service market such as Google. The data-driven origin of those companies could provide them with a competitive edge with regard to the customisation of services, however. Since the business model originates in digital, data-driven markets, companies such as Google or Facebook might already be in possession of data on customer preferences and buying behaviour. Hence, a more accurate personalisation of services could be expected. It may further be stated that the approach of MaaS as proposed by Finnish Transport Agency (2015), Heikkilä (2014) and Hietanen (2014, 2 f.) to a certain extent is based on this business model.

Furthermore, it could be claimed that there are various viabilities to create *hybrid versions* of the two previous business model archetypes. van Audenhove et al.

(2014, p. 23) discuss the combination of both models to a 'total mobility provider' which complements its integrated transport services by aggregating additional services such as car park reservations or taxi services. Notwithstanding, what is neglected by the authors is the diversification of aggregators investing in physical assets or operating transport services under a franchise model, hence, partially transforming into integrators, which should be seen as a feasible hybrid version as well.

Another archetype mostly advocated by van Audenhove et al. (2014) and Lerner and van Audenhove (2012) focuses on the provision of complete mobility system solutions for a specific region. The concept which pursues a business-to-business (B2B) approach offers all components required for the operation of a mobility system including infrastructure, vehicles and traffic management. It could be claimed that this specialist approach demonstrates significant similarities with the business model of current transport associations.

Simultaneously, the question of companies that could potentially take over the role of such an operator arises. Rossbach et al. (2013, pp. 33–37) analysed the suitability of companies from several industries with regard to their core competencies to act as an orchestrator for mobility chains. Based on a qualitative criteria catalogue the authors conclude that trans-regional transport service providers might be the most suitable to act as a HMP, whereas car manufacturers are the most advanced in terms of technology required. The analysis is summarised in Table 3.2. Nevertheless, Rossbach et al. (2013) disregarded other potential operators such as governmental institutions which are suggested by Aberle and Werbeck (2013, p. 54). Thus, the table was complemented by the author with further potential operators.

Table 3.2 Potential holistic mobility service providers

Type of company	Main rationales for suitability	Business origin
Car manufacturers	Already active members of the mobility industry, partially with digital mobility services (e.g. car sharing); strong competence in managing alliances and networks; strong brands; experience in necessary technology as well as strong R&D activities	Integrator/ asset driven
Transport associations	Already active members of the mobility industry, partially with digital mobility services (e.g. car sharing); dominant role and quasi monopolists in their region	Integrator/ asset driven
Railway companies	Already active members of the mobility industry, partially with digital mobility services (e.g. car sharing); specifically strong on long-distance trips; oligopolistic or rather monopolistic market structure makes such companies hard to replace; strong brands	Integrator/ asset driven
Airlines	Already active members of the mobility industry; specifically strong on long-distance trips; oligopolistic or rather monopolistic market structure makes such companies hard to replace; strong brands	Integrator/ asset driven
Telecommunications providers	In possession of infrastructure required for connectivity/Internet of Things; strong in end-customer management; ability to deliver, suppress and cash up large-scale mass services transactions	Aggregator/ data driven
Internet companies	Short time-to-market; high technology competence; large reach of customers; experience in aggregating third party services; data on customer behaviour; usage of geo-commerce	Aggregator/ data driven
Technology firms	Experience in system integration; access to customer-specific data; partially strong brands (e.g. Apple); strong capabilities in R&D; strong in end-customer management	Aggregator/ data driven
Public institutions	Cooperation and participation of 'sub'-contractors can be enforced by law; public services could appeal more trustful to customers	Integrator/ asset driven
Trading companies	Strong brands; access to some information on customer demand and requirements	Aggregator/ data driven

Source: Own depiction

References

Aberle, C., & Werbeck, I. (2013). *Komfortabel, vernetzt, klimafreundlich: Online-Mobilitätsangebote als Instrumente einer Nachhaltigen Entwicklung? Studie in Auftrag des Rates für Nachhaltige Entwicklung.* Rat für Nachhaltige Entwicklung.

Accenture. (2014). *Customer 2020: Are you future-ready or reliving the past?: Ten years of Accenture research highlights real opportunities for providers to better meet customers' steadily rising expectations* (No. 14-6613). Accenture.

Accenture. (2016). *Customer centricity: Create true customer centricity by understanding customers' "Digital Intensity"*. Retrieved August 10, 2021, from https://www.accenture.com/us-en/insight-create-true-customer-centricity-understanding-customers

Allen, J., Reichheld, F. F., Hamilton, B., & Markey, R. (2005). *Closing the delivery gap: How to achieve true customer-led growth*. Retrieved August 20, 2021, from http://www.bain.com/publications/articles/closing-the-delivery-gap-newsletter.aspx

Bechmann, R., Scherk, M., Heimann, R., & Schäfer, R. (2015). *Trendanalyse: Vernetztes Fahrzeug 2015: Die wichtigsten Trends und Herausforderungen in der Fahrzeugtelematik*. MBtech Consulting.

Berry, L. L., & Carbone, L. P. (2007). Build loyalty through experience management. *Qualilty Progress, 40*(9), 26–32.

Berry, L. L., Carbone, L. P., & Haeckel, S. H. (2002). Managing the total customer experience. *MIT Sloan Management Review, 43*(3), Reprint.

Chen, I. J., & Popovich, K. (2003). Understanding customer relationship management (CRM). *Business Process Management Journal, 9*(5), 672–688. https://doi.org/10.1108/14637150310496758

Conroy, P., Porter, K., Nanda, R., Renner, B., & Narula, A. (2015). *Consumer products trends: Navigating 2020*. Deloitte University Press.

Dyché, J. (2001). *The CRM handbook: A business guide to customer relationship management*. Addison-Wesley.

Finnish Transport Agency. (2015). *MaaS services and business opportunities* (Research reports of the Finnish transport agency no. 56/2015). Finnish Transport Agency.

HBR Analytical Services & Strativity. (2015). *Making customer-centric strategies take hold*. Harvard Business School Publishing.

Heikkilä, S. (2014). *Mobility as a service – A proposal for action for the public administration. Case Helsinki* (Master's thesis). Aalto University, Espoo.

Hietanen, S. (2014). 'Mobility as a service' - The new transport model? *Eurotransport, 12*(2), 2–4.

Homburg, C., Jozić, D., & Kuehnl, C. (2013). *Customer experience management* (IMU research insights No. 019). University of Mannheim.

Homburg, C., Jozić, D., & Kuehnl, C. (2015). Customer experience management: Toward implementing an evolving marketing concept. *Journal of the Academy of Marketing Science.* https://doi.org/10.1007/s11747-015-0460-7.

Kamaldevi, B. (2010). Customer experience management in retailing. *Business Intelligence Journal, 3*(1), 37–54. https://doi.org/10.1016/j.jretai.2009.01.001

Kuhnimhof, T., Buehler, R., Wirtz, M., & Kalinowska, D. (2012). Travel trends among young adults in Germany: Increasing multimodality and declining car use for men. *Journal of Transport Geography, 24*, 443–450. https://doi.org/10.1016/j.jtrangeo.2012.04.018

Lerner, W., & van Audenhove, F.-J. (2012). Die Zukunft der städtischen Mobilität - auf dem Weg zu vernetzten, multimodalen Städten im Jahr 2050. *PTI Magazine., 3*, 14–18.

Leußer, W., Hippner, H., & Wilde, K. D. (2011). CRM – Grundlagen, Konzepte und Prozesse. In H. Hippner, B. Hubrich, & K. D. Wilde (Eds.), *Grundlagen des CRM. Strategie, Geschäftsprozesse und IT-Unterstützung* (3rd ed., pp. 15–56). Gabler Verlag.

Malthouse, E. C., Haenlein, M., Skiera, B., Wege, E., & Zhang, M. (2013). Managing customer relationships in the social media era: Introducing the social CRM house. *Journal of Interactive Marketing, 27*(4), 270–280. https://doi.org/10.1016/j.intmar.2013.09.008

Meyer, C., & Schwager, A. (2007). Understanding customer experience. *Harvard Business Review.* (Reprint R0702G). Retrieved from http://harvardbusinessonline.hbsp.harvard.edu/relay.jhtml?name=itemdetail&referral=4320&id=R0702G

Parment, A. (2013). *Die Generation Y: Mitarbeiter der Zukunft motivieren, integrieren, führen* (2nd ed.). Springer Fachmedien.

Payne, A. (2005). *Handbook of CRM: Achieving excellence in customer management*. Butterworth-Heinemann.

Payne, A., & Frow, P. (2005). A strategic framework for customer relationship management. *Journal of Marketing, 69*(4), 167–176. https://doi.org/10.1509/jmkg.2005.69.4.167

PwC. (2016). *Global total retail survey 2016: Key finding.* Retrieved May 17, 2021, from http://www.pwc.com/gx/en/industries/retail-consumer/global-total-retail.html

Rammler, S., & Sauter-Servaes, T. (2013). *Innovative Mobilitätsdienstleistungen* (Arbeitspapier No. 274). Hans-Böckler-Stiftung.

Rosman, R., & Stuhura, K. (2013). The implications of social media on customer relationship management and the hospitality industry. *Journal of Management Policy and Practice v, 14*(3), 18–26.

Rossbach, C., Winterhoff, M., Reinhold, T., Boekels, P., & Remane, G. (2013). *Connected Mobility 2025: Neue Wertschöpfung im Personenverkehr der Zukunft* (Think:act Study. In-depth knowledge for decision makers). Roland Berger Strategy Consultants.

Siegfried, P. (2014). *Knowledge transfer in service research - Service engineering in startup companies.* EUL-Verlag.

Ternès, A., Towers, I., & Jerusel, M. (2015). *Konsumentenverhalten im Zeitalter der Mass Customization: Trends: Individualisierung und Nachhaltigkeit.* Springer Fachmedien.

van Audenhove, F.-J., Korniichuk, O., Dauby, L., & Pourbaix, J. (2014). *The future of urban mobility 2.0: Imperatives to shape extended mobility ecosystems of tomorrow.* Retrieved from Arthur D. Little/The International Association of Public Transport (UITP) website: www.adl.com/FUM2.0

Vandermerwe, S., & Rada, J. (1988). Servitization of business: Adding value by adding services. *European Management Journal, 6*(4), 314–324.

Verhoef, P. C., Lemon, K. N., Parasuraman, A., Roggeveen, A., Tsiros, M., & Schlesinger, L. A. (2009). Customer experience creation: Determinants, dynamics and management strategies. *Journal of Retailing, 85*(1), 31–41. https://doi.org/10.1016/j.jretai.2008.11.001

Winer, R. S. (2001). A framework for customer relationship management. *California Management Review, 43*(4), 89–105.

Winterhoff, M., Kahner, C., Ulrich, C., Sayler, P., & Wenzel, E. (2009). *Zukunft der Mobilität 2020: Die Automobilindustrie im Umbruch?* Arthur D. Little.

Methodology of the Survey-Based Expert Analyses

4

4.1 Research Questions

During the literature review outlined in the previous chapters, several gaps in the existing literature were identified which should be analysed in order to answer the overarching problem definition of this research.

The mobility behaviour of persons has been subject to thorough research. The impact and usage of information and ICT were discussed amongst others by Otto (2010, p. 2) and Pawlak et al. (2015, p. 31). It was concluded that the availability of decision-relevant information has the potential to alter current mobility habits. Moreover, it was found out that ICT is already utilised in the transport mode selection processes. The type of information required at every stage of a trip was analysed as well by Grotenhuis et al. (2007). The extent or the manner ICT and information is utilised in everyday mobility decisions has not been analysed in greater detail, however.

Besides, the implementation of ADV, the role of private cars and shared services in future mobility systems was discussed extensively. Nevertheless, the attractiveness of ADV for different use cases should be evaluated from a consumer's point of view. This may be largely due to the cultural barrier towards ADV as advocated by Kröger (2015, p. 64), which requires a practical verification. Furthermore, previous studies virtually agree upon customer-centric and personalised mobility service provision (e.g. Finnish Transport Agency, 2015; Kamargianni et al., 2015; Fishman, 2012; Capgemini, 2013). As mentioned at the beginning of this research, customer centricity may call for a shift from a plain mobility management to a CCM perspective. Nevertheless, the acceptance of CCM practices, especially CRM and the integration of non-mobility VAS, has not been analysed properly in the context of (everyday) mobility. The role of CEM in everyday mobility has not been analysed, yet. Therefore, current practices in everyday mobility as well as the identification of (future) requirements that might be placed on mobility service providers should be addressed.

© The Author(s), under exclusive license to Springer Nature Switzerland AG 2022
P. Siegfried, *Digitalisation in Mobility Service Industry*, Future of Business and
Finance, https://doi.org/10.1007/978-3-031-07151-5_4

Business model archetypes and companies which may serve as mobility chain orchestrators have been pointed out by previous studies as well. Subsequently, it should be evaluated which business model is the most preferred from a user's point of view. Moreover, since companies differ in their suitability for applying the respective business models it seems crucial to evaluate which firm is seen as the most capable to act as a HMP. Especially the competition between asset-driven (i.e. traditional mobility service providers) and data-driven (i.e. new entrants in the mobility market such as Internet firms) companies should be analysed in greater detail. As pointed out by Rossbach et al. (2013, p. 26) consumers already rate their user experiences made with companies such as Google or Apple higher than with for instance automotive original equipment manufacturers (OEM). The authors also claim that convenience in the usage of services and customer experience might become more important than the brand of a vehicle or operator. This research thus should also be concerned with the evaluation of a potential shift in the dominant players in the mobility service market.

Finally, it should be found out how future digital mobility markets will be structured. To achieve this, insights gathered during this research should be utilised to create potential scenarios. To conclude, this research specifically tries to answer the following research questions:

1. How does everyday mobility in the age of digitalisation look like and what requirements can be derived thereof for mobility services providers?
2. Will mobility management transform into a holistic customer-centric management that also includes non-mobility services?
3. Will traditional mobility service providers revert into mere transport operators and data-driven companies take over the role as holistic mobility chain orchestrators?
4. What are possible scenarios for a digitalised mobility market?

4.2   Selection of Research Method and Instrument

The selection of a research project's methodology is dependent on various factors such as the current state of knowledge in the respective field and the research's overall orientation. By recapitulating the purpose of this research it was decided to select a quantitative approach. This has several reasons. First of all, in order to verify assumptions or affirm theories quantitative research methods should be applied (cf. Weathington et al., 2012, p. 398). Since this research follows the assumption that digitalisation may cause a shift of paradigm in both the dominant market players as well as in the interaction between service providers and their customers, it could be seen as theory testing. In addition, through the utilisation of standardised instruments quantitative methods enable a researcher to obtain data suitable to make predictions (cf. Harwell, 2011, p. 149). This may be especially important if the development of a quantification model as well as the development of scenarios for future mobility market is considered. Moreover, this research intends to verify general assumptions about a potential paradigm change in the most objective manner possible. Hence, in-depth studies considering subjective experiences and perspectives as applied in the qualitative tradition (cf. Harwell, 2011, p. 148) may interfere with this objective.

Regarding the selection of research instrument it was decided to conduct survey-based research, which is one of the most popular data collection techniques in the field of social and behavioural sciences (cf. Royal et al., 2010, p. 609; Neuman, 2014, p. 316). More specifically it was opted for a standardised survey design in form of a web-based questionnaire. The rationale behind this may be manifold. First of all, as pointed out by Kelley et al. (2003, p. 262) survey research enables a researcher to generate a relatively large set of data with a broad bandwidth of individuals at a considerably low cost. Moreover, a standardised survey is highly suitable for the collection of data as objectively and uniformly as possible (cf. Reinecke, 2014, p. 603), which is in line with the goals of this research. The conduct of a web-based survey provides additional advantages. Through the distri-bution, online administrative costs as well as the time questionnaires are received by the participants decrease significantly (cf. Evans & Mathur, 2005, 197 f.). In addition, the data obtained is already in a digital format which reduces the risk of data distortion caused by manual conversion. Besides, the order questions are answered as well as the design of specific question order logics may be solely possible if the questionnaire is distributed digitally (cf. Evans & Mathur, 2005, p. 200).

4.3 Design of Research Instrument

Since questionnaire designs have significant impacts on response rates and quality special attention was paid to previous studies that focused on the minimisation of negative effects before designing the actual questionnaire. As 'conventional wis-dom' Krosnick and Presser (2010, p. 264) state the usage of simple, concrete syntax and words as well as the avoidance of words with ambiguous meanings. In addition, the response options to a question should be exhaustive and questions on the same topic should be grouped together. Eichhorn (2004, p. 18) raises the importance of a well-structured introduction to motivate and persuade potential participants to engage in filling out the questionnaire.

An issue commonly observed in the conduct of questionnaire-based surveys is the generation of sufficient responses (cf. Sax et al., 2003) as well as the retention of persons to complete a survey. Ganassali (2008, p. 23) points out that longer questionnaires may yield fewer responses and suggests an optimal length ranging between 15 and 30 questions. In addition, the author mentions the importance of a point of completion (POC) indicator. Thielsch and Weltzin (2009, p. 72) advocate for a limited usage of open questions as they are suspected to cause high dropout rates. In their research Frick et al. (1999, p. 213) found out that the announcement of a lottery at the beginning could reduce dropout rates by up to 50%. Eichhorn (2004, p. 20) agrees upon the importance of incentives. He claims that even non-financial

incentives are a good method to convey the feeling of providing a valuable contribution to the survey participants. Even though Göritz (2004, p. 341) claims that
incentives do not have an impact on response quantity and data quality, the provision
of a few gifts is recommended as standard incentives.

Thielsch and Weltzin (2009, p. 73) also underline the importance of a high degree
of usability in web-based questionnaires. The authors state that besides clearly
formulated items and instructions the survey should be accessible from a broad
range of devices. By considering the increasing utilisation of mobile devices it
should be claimed that the usability of questionnaires on those devices may be a
prerequisite for achieving a sufficient response rate.

After a thorough revision of relevant literature, the questionnaire for this research
was designed in the following manner. In an iterative process, the number of
questions was reduced from initially 24 to 11 in order to minimise the risk of early
dropouts due to a too long questionnaire. The syntax of the selected questions was
altered several times to achieve the highest simplification and most appropriate
response options possible as suggested by Krosnick and Presser (2010, p. 264).
The 11 final questions were then assigned to three categories which should also
represent the questionnaire's main categories. Subsequently, the design was tested in
a pretest research where 267 valid interviews were collected.[1] The results of the
pretest were utilised to alter the survey design at several places where the questionnaire yielded insufficient results. This questionnaire enhancement resulted in a slight
increase in questions from 11 to a maximum of 15. The number of questions
depended on different question logic orders. Open questions were solely utilised
as follow-up questions and limited to three. The questionnaire's main categories
remained unchanged.

At the beginning of the survey, the participant was shown an introductory page.
The text on this page was formulated to motivate the participant to start the survey.
To achieve this, provocative assertions which—at first glance—seemed utopic were
stated. Furthermore, the importance of the participant's opinion as well as the
assurance of anonymity and privacy was stated. As recommended by Frick et al.
(1999, p. 213), Göritz (2004, p. 341) and Eichhorn (2004, p. 20), a lottery for several
technological gadgets provided by Messe Frankfurt was announced. Moreover, the
option to claim the survey results free of charge was mentioned. The usage of
incentives was affirmed by the pretest results as well, where around 72% of the
participants signed in for the lottery and/or were interested in the research's results.
A POC indicator was placed on the lower right-hand side as well.

After the introduction, the first set of questions was shown which consisted of up
to six questions regarding the participants' current mobility behaviour which served
as 'icebreaker questions' to let the participants accustom themselves to the survey's

---

[1] The pretest was carried out between the 1st and 30th of May 2016.

topic. Afterwards four questions concerning the attractiveness of CCM in mobility services were asked. This included the verification of potential VAS outlined in Sect. 2.4.4 as representation for such services in general. The third category comprised questions regarding the attractiveness of the business models outlined in Sect. 3.3 as well as general preferences associated with the respective business models. The business models were expanded by a non-digital model in order to enhance the depiction of the aspired mobility split. Moreover, the participant should indicate which type of companies in his or her opinion are the most suitable for tasks and components of the respective business models. In the end, the participant was asked to provide information on his or her age, occupancy as well as the post-code of his or her primary residence.

4.3.1 Scale Selection and Design

The selection and design of the most appropriate rating scale are more complex than it initially appears. Therefore, the issue received significant attention from academic researchers. While the selection of scales for questions with a nominal character might be less complicated, the design of scales for questions that should analyse opinions and viewpoints can be seen as an art and science alike (cf. Royal et al., 2010, p. 609). Thus, the following paragraphs should focus on the design of those scales.

The most commonly used scale was introduced by Likert in 1932 and therefore is referred to as the Likert Scale (cf. Royal et al., 2010, p. 609). Due to its numerous validations in the application in survey research, it was decided to utilise a Likert-type scale for the applicable questions in this survey.

During the literature review, three main areas which should be considered in the design of a Likert Scale were identified: the number of elements, the usage of a midpoint, and the labelling of the scale. Various studies have analysed the effects of variations in the number of elements on a Likert Scale and the quality of responses. In an extensive literature review Menold and Bogner (2015, p. 2) conclude that even though Likert scales could contain up to 100 elements, a range between five and seven elements should be preferred. Krosnick and Presser (2010, p. 272) in addition highlight that while the reliability increased significantly from 2- to 3- to 5-point scales, the increase after seven points were insignificant.

Another factor that was analysed extensively is the inclusion of so-called midpoints (i.e. evenly numbered versus oddly numbered scales). There are two contrasting views found in the respective literature. On the one hand, some authors argue that the existence of a midpoint could cause participants to be less thorough in their response selection as well as provoke a stronger tendency to choose the midpoint (cf. Menold & Bogner, 2015, p. 5). This phenomenon is referred to as 'satisficing' and may be highly dependent on the motivation to provide accurate responses as well as the cognitive skills of a participant (cf. Krosnick & Presser, 2010, p. 271). On the other hand, it is argued that some of the participants might be genuinely indecisive on the presented opinion and thus are forced to 'choose sides' (cf. Menold & Bogner, 2015, p. 5; Royal et al., 2010, p. 610; Krosnick & Presser, 2010, p.271). Royal et al. (2010, p. 610) stress the importance of scale length in this regard. Accordingly, reliability is compromised in shorter scales when a midpoint is utilised. The definition of a 'shorter scale' is, however, not further specified.

The labelling of scales has a significant influence on data quality as well. In general three options are discussed: full verbal labelling, verbal labelling of extremes and numeric labelling. Royal et al. (2010, p. 609) and Menold and Bogner (2015, p. 2) amongst a broad range of studies advocate for a full verbal labelling as the measurement of validity is considered to improve. Moreover, 'direct labelling' should be applied (i.e. adapting a scale's labelling to the context of the question) in order to avoid the tendency of participants to select the 'agree' option (cf. Menold & Bogner, 2015, p. 8).

The scale design for this research's questionnaire has been determined as follows. It was decided to limit the scale points of the concerned questions to four and five, respectively. The main rationale behind this is that with the increase in scale points the cognitive effort increases as well (cf. Krosnick & Presser, 2010, p. 271). Due to the futuristic orientation of the questionnaire, which may require vast cognitive effort, it was decided to reduce the effort in response selection. For questions with a five-point scale, the fifth point represented the feasibility to waive the question. A conventional mid-point was avoided. This decision was founded on the pretest results where a strong tendency towards the midpoint was observed in those questions. The scales were fully labelled and adapted to the context of the respective question (i.e. direct labelling).

4.3.2 Consideration of Further Potential Biases

Besides the design of scales, the consideration of other biases influencing the results seems crucial to assure high-quality responses. Krosnick and Presser (2010) identify the failure to properly recall events from the past (recall error) as one major bias. Other biases are mostly related to the deviation of the selected response options and the actual opinion of a survey participant. Moors et al. (2014) mention two response biases in this respect. As extreme response style (ERS) the authors define a participant's tendency to only select the extreme endpoints on a scale. Acquiescence response style (ARS) on the other hand refers to the tendency to agree with statements presented regardless of an item's content (cf. Moors et al., 2014, p. 370). Krosnick and Presser (2010, p. 285) also state that respondents may select items according to their social desirability or undesirability (social desirability response bias, SDRB).

Another source of potential biases is the static presentation of items. Several studies mention the so-called 'primacy effect' where options are more likely to be selected when they are presented early (cf. Krosnick & Presser, 2010, p. 278; Eichhorn, 2004, p. 23; Menold & Bogner, 2015, p. 4). A special form of this phenomenon is referred to as 'general primacy effect' which describes the tendency to select items that are located on the left-hand side of a scale (cf. Menold & Bogner, 2015, p. 4).

In order to mitigate the above-mentioned biases several preventive measures have been implemented into the questionnaire design. To avoid the primacy effect the available items to a question were automatically randomized for every participant. Moreover, the questions were asked in a manner in which the recall of past events was made obsolete to reduce the effects of recall errors. Nonetheless, it may be argued that it is impossible to mitigate the risk of every bias. This may be due to the fact that response styles such as ARS, ERS and SDRB are highly dependent on the individual's motivation to genuinely contribute to the survey.

It should further be stated that the questionnaire design stages are not fully independent from each other. Thus, some potential biases have already been taken into account in previous stages. Eichhorn (2004) for instance recognises the tendency to select the midpoint of scale as a potential bias or rather a source of error. Notwithstanding, this risk may be already mitigated through the scale design utilised in the questionnaire. Krosnick and Presser (2010) mention question order effects that affect response quality. Nonetheless, the described serial order effect (i.e. impact of the question order on the participants' motivation) and semantic order effect (i.e. disturbance of cognitive processing caused by the question order) have been taken into account in previous design stages by arranging the questions according to guidelines suggested by the authors. Furthermore, a slight moderation of ARS effects may be achieved through direct labelling of scales. SDRB is expected to decrease if the level of anonymity in a survey increases (cf. Krosnick & Presser, 2010, p. 286). Since the survey is conducted as a web-based questionnaire a mitigation of SDRB thus might be expected.

4.4   Sample Selection

Due to the cost intenseness of a full census, most empirical studies are based on samples which should represent the target population (cf. Häder & Häder, 2014, p. 283). Sample selection methods can broadly be divided into two categories. *Probability sampling methods* are based on probabilistic mechanisms that should give every member of the target population a certain likelihood to be selected into the sample (cf. Fricker, 2008, p. 199). This probability must be larger than zero (cf. Skowronek & Duerr, 2009, p. 412). *Non-probability sampling methods* on the other hand rely on techniques that disregard the equality of selection, but rather select a sample on the basis of availability or personal choice (cf. Fricker, 2008, p. 199). The degree of randomness in the sampling process has implications on the potential to infer from research findings to the target population. It is argued that due to the negligence of giving every individual a certain probability to be selected findings drawn from non-randomised samples are unsuitable to be generalised (cf. Cooper & Greenaway, 2015, p. 2). Notwithstanding, the selection of a specific sampling method should be in accordance with a research project's objectives as well as resources and time frame (cf. Cooper & Greenaway, 2015, p. 11).

This research applied non-probability sampling in terms of convenience sampling. The decision can be traced back to several factors. Firstly, this research is subject to contract research which on the one hand provided the researcher with the opportunity to gain access to a large survey panel, but on the other hand quasi predetermined the sampling selection method. Secondly, the view on probabilistic techniques as universal sampling solution is increasingly challenged, as there is no existing research that truly satisfies the characteristics of 'textbook probability sampling' (cf. Doherty, 1994, p. 23; Wretman, 2010, p. 30) and the increasing cost intenseness of those methods (cf. Baker et al., 2013, p. 6). This might be specifically important if the limited timeframe and budget of this research is considered. Moreover, non-random sampling is highly suitable to evaluate or establish the existence of a research problem (cf. Skowronek & Duerr, 2009, p. 413) which is in accordance with the aim of this research.

As mentioned before, statistical inference of non-probability samples is seen as challenging. Nevertheless, Skowronek and Duerr (2009, p. 413) suggest measures to enhance the usefulness of data generated under such conditions. As one measure the authors propose to select the accessible sample according to characteristics of the target population in terms of socio-demographic factors. Another measure is seen in increasing the sample's diversity by distributing the questionnaire at different places and times. The third measure suggested is to increase the sample size. The respective sample size requirements should be determined with methods applied in probabilistic sampling (cf. Skowronek & Duerr, 2009, p. 413). These guidelines were applied in the sample selection process as described in the following paragraph.

The sample recruitment was conducted utilising active and passive sampling techniques. Since there was no information available on socio-demographic factors it was waived to conduct a preselection of participants. The survey invitation was distributed to a panel of more than 14,500 potential participants which were derived from three sources in order to maximise diversity and the sample size as a whole. The largest source was the newsletter database of Messe Frankfurt which comprised 14,280 email addresses. This database consists of exhibitors and fair visitors which subscribed to a newsletter that is concerned with digital trends and novelties. The second source was the email distribution list of the researcher's university that consisted of email addresses of students, academic as well as administrative staff. The third source was an email list generated during the pretest phase of this research. For passive sampling, the questionnaire link was placed on websites as well as social media presences of Messe Frankfurt and supporting websites (see e.g. Schmidt, 2016). In addition, snowball sampling was applied: in the email invitation participants were asked to redistribute the questionnaire link in their social environment. This should enhance the diversity and overall sample size as well.

4.5   Data Collection and Analysis Approach

The survey was conducted using the online survey and software provider SoSci Survey developed by Leiner (2016). The provider was selected for three main reasons. First of all, SoSci Survey provides the feasibility to programme individual filters and create various logical question orders by offering the questionnaire's source code. This also enables a complete customisation of the questionnaire design, which was important to create a template satisfying the requirements on corporate identity of Messe Frankfurt. In addition, SoSci Survey possesses a rating scheme to evaluate the quality of every finished interview. The programme rates an interview on two aspects and assigns a total malus score which indicates the quality of the data set. The malus score is calculated on the basis of left out questions as well as time spent on answering the questionnaire. Furthermore, the programme offers a large variety of functions to manage the survey panel such as individual email invitations and reminders.

The data collection or rather the questionnaire distribution was organised in waves to improve the sample diversity. As field time the period between the 13th of July and the 22nd of August was determined. In the first wave the survey invitation was distributed to the researcher's university's email distribution list. Simultaneously, the invitation link was placed on some social media presences of Messe Frankfurt. In a second wave the pretest research participants were invited to take place in the main survey. In a third wave the invitation was distributed to the newsletter database of Messe Frankfurt. Throughout the second and third waves reminders where sent through the social media accounts of Messe Frankfurt. In addition, the survey was promoted by a guest blog post of the researcher's university's logistics institute.[2]

Due to the time constraints this research was subject to the database for the analysis presented in the following paragraphs does not represent the full data set gathered throughout the research, but rather an analysis of interim results. As due date for the data collection the 30th of July 2016 was determined. At this stage, the third source of potential participants (i.e. the newsletter database of Messe Frankfurt) had not been invited to participate in the research. The gathered data was analysed with descriptive and explorative statistics using IBM's Statistical Package for the Social Sciences (SPSS).

References

Capgemini. (2013). *Mobilität der Zukunft: Reisen wird zum entspannten Erlebnis durch intermodale, integrierte und digitale Lösungen.* Capgemini.

Cooper, D., & Greenaway, M. (2015). *Non-probability survey sampling in official statistics* (ONS methodology working paper series No. 4). United Kingdom Office for National Statistics.

---

[2]See Benz and Jonas (2016).

Doherty, M. (1994). Probability versus non-probability sampling in sample surveys. *The New Zealand Statistics Review*. (March 1994 Issue), 21–28.

Eichhorn, W. (2004). *Online-Befragung: Methodische Grundlagen, Problemfelder, praktische Durchführung*. Retrieved from http://wolfgang-eichhorn.com/cc/onlinebefragung-rev1.0.pdf

Evans, J. R., & Mathur, A. (2005). The value of online surveys. *Internet Research, 15*(2), 195–219. https://doi.org/10.1108/10662240510590360

Finnish Transport Agency. (2015). *MaaS services and business opportunities* (Research reports of the Finnish transport agency no. 56/2015). Finnish Transport Agency.

Fishman, T. D. (2012). *Digital-age transportation: The future of urban mobility*. Deloitte University Press.

Frick, A., Bächtiger, M. T., & Reips, U.-D. (1999). Financial incentives, personal information and dropout rate in online studies. *Dimensions of Internet science*, 209–219.

Fricker, R. D. (2008). Sampling methods for web and e-mail surveys. In N. Fielding, R. M. Lee, & G. Blank (Eds.), *The SAGE handbook of online research methods* (pp. 195–216). Sage Publications.

Ganassali, S. (2008). The influence of the design of web survey questionnaires on the quality of responses. *Survey Research Methods, 2*(1), 21–33.

Göritz, A. S. (2004). The impact of material incentives on response quantity, response quality, sample composition, survey outcome, and cost in online access panels. *International Journal of Market Research, 46*(3), 327–345.

Grotenhuis, J.-W., Wiegmans, B. W., & Rietveld, P. (2007). The desired quality of integrated multimodal travel information in public transport: Customer needs for time and effort savings. *Transport Policy, 14*(1), 27–38. https://doi.org/10.1016/j.tranpol.2006.07.001

Häder, M., & Häder, S. (2014). Stichprobenziehung in der quantitativen Sozialforschung. In N. Baur & J. Blasius (Eds.), *Handbuch Moethoden der empirischen Sozialforschung* (1st ed., pp. 283–297). Springer Fachmedien.

Harwell, M. R. (2011). Research design in qualitative/quantitative/mixed methods. In C. F. Conrad & R. C. Serlin (Eds.), *The SAGE handbook for research in education. Pursuing ideas as the keystone of exemplary inquiry* (2nd ed., pp. 147–163). Sage Publications.

Kamargianni, M., Matyas, M., Li, W., & Schäfer, A. (2015). *Feasibility study for "mobility as a service" concept in London: FS-MaaS project – Final deliverable*. UCL Energy Institute.

Kelley, K., Clark, B., Brown, V., & Sitzia, J. (2003). Good practice in the conduct and reporting of survey research. *International Journal for Quality in Health Care, 15*(3), 261–266. https://doi.org/10.1093/intqhc/mzg031

Kröger, F. (2015). Das automatisierte Fahren im gesellschaftsgeschichtlichen und kulturwissenschaftlichen Kontext. In M. Maurer (Ed.), *Autonomes Fahren. Technische, rechtliche und gesellschaftliche Aspekte* (pp. 41–67). Springer.

Krosnick, J. A., & Presser, P. (2010). Question and questionnaire design. In P. V. Marsden & J. D. Wright (Eds.), *Handbook of survey research* (2nd ed., pp. 263–313). Emerald.

Leiner, D. J. (2016). *SoSci survey version 2.6.00-i*. Retrieved from https://www.soscisurvey.de

Menold, N., & Bogner, K. (2015). *Gestaltung von Ratingskalen in Fragebögen* (GESIS Survey Guidelines). GESIS – Leibniz-Institut für Sozialwissenschaften.

Moors, G., Kieruj, N. D., & Vermunt, J. K. (2014). The effect of labeling and numbering of response scales on the likelihood of response bias. *Sociological Methodology, 44*(1), 369–399. https://doi.org/10.1177/0081175013516114

Neuman, W. L. (2014). *Social research methods: Qualitative and quantitative approaches* (Pearson new international edition). Pearson.

Otto, S. (2010). *The psychology of transport choice*. Institute for Ecological Economic Research (IÖW).

Pawlak, J., Le Vine, S., Polak, J., Sivakumar, A., & Kopp, J. (2015). *ICT and physical mobility: State of knowledge and future outlook*. Institute for Mobility Research (ifmo).

Reinecke, J. (2014). Grundlagen der standardisierten Befragung. In N. Baur & J. Blasius (Eds.), *Handbuch Moethoden der empirischen Sozialforschung* (1st ed., pp. 601–617). Springer Fachmedien.

Rossbach, C., Winterhoff, M., Reinhold, T., Boekels, P., & Remane, G. (2013). *Connected Mobility 2025: Neue Wertschöpfung im Personenverkehr der Zukunft* (Think:act study. In-depth knowledge for decision makers). Roland Berger Strategy Consultants.

Royal, K. D., Ellis, A., Ensslen, A., & Homan, A. (2010). Rating scale optimization in survey research: An application of the Rasch rating scale model. *Journal of Applied Quantitative Methods, 5*(4), 607–617.

Sax, L. J., Gilmartin, S. K., & Bryant, A. N. (2003). Assessing response rates and nonresponse bias in web and paper surveys. *Research in Higher Education, 44*(4), 409–432. https://doi.org/10.1023/A:1024232915870

Schmidt, E. (2016). *Umfrage: Sterben Daimler & Co. bald aus?: Umfrage zur Akzeptanz von Mobilitätsdienstleistungen.* Retrieved July 27, 2021, from http://www.kfz-betrieb.vogel.de/umfrage-sterben-daimler-co-bald-aus-a-542756/

Skowronek, D., & Duerr, L. (2009). The convenience of nonprobability: Survey strategies for small academic libraries. *College & Research Libraries News, 70*(7), 412–415.

Thielsch, M. T., & Weltzin, S. (2009). Online-Befragungen in der Praxis. In T. Brandenburg & M. T. Thielsch (Eds.), *er Praxis. Praxis der Wirtschaftspsychologie: Themen und Fallbeispiele für Studium und Praxis* (pp. 69–85). MV Wissenschaft.

Weathington, B. L., Cunningham, C. J. L., & Pittenger, D. J. (2012). *Understanding business research.* Wiley.

Wretman, J. (2010). Reflections on probability vs nonprobability sampling. In M. Carlson, H. Nyquist, & M. Villani (Eds.), *Official statistics – Methodology and applications in honour of Daniel Thorburn* (pp. 29–35). Department of Statistics, Stockholm University.

Empirical Findings from the Survey-Based Expert Analyses

5

5.1 Sample Attributes

During the field time, a total of 460 fully completed interviews were generated. The retention rate[1] of all participants was 81.85% which should be regarded as an acceptable ratio if the questionnaire's difficulty is considered. The overall response quality can be seen as relatively high. Out of the 460 completed interviews merely four had demonstrated degraded scores above 100, which is suggested as the threshold for poor data quality in the rating scheme of SoSci Survey. Therefore, merely 0.0087% of the completed interviews had to be excluded from the analysis.

Regarding the participants' age, a strong manifestation in the range of 40–50 years was found. The average age is 41.34 years while the mode lies in the category of 45–54 years which make up 30.9% of the total participants. In order to enhance the analysis across different generations, the participants were categorised accordingly. The categorisation scheme was derived from Nicholas (2009, 47 f.). According to this categorisation, the largest share of participants belong to 'Generation X' (39.9%), followed by 'Generation Y' (35.7%) and 'Baby Boomers' (24.1%). The 'Traditionalists' are represented marginally with 0.2% whereas no responses from 'Generation Z' were obtained.

The strongest manifestations concerning the community size is found in the category of cities with more than 500,000 inhabitants (29.4%). Moreover, nearly half of the respondents (49.9%) indicated not to have children. It should also be stated that the educational level of the survey panel is relatively high and above average. The most reported educational degrees are the German university diploma (29.1%), bachelor's degree (10.8%), master's degree (8.9%) and doctoral degree

---

[1] As retention rate the ratio of participants who completed the questionnaire compared to the total amount of participants including those who aborted the survey is understood.

© The Author(s), under exclusive license to Springer Nature Switzerland AG 2022
P. Siegfried, *Digitalisation in Mobility Service Industry*, Future of Business and Finance, https://doi.org/10.1007/978-3-031-07151-5_5

(7.3%). In total, 60.18% of the valid responses to this question are in possession of a university degree while merely 0.7% indicated the educational level of the German 'Hauptschule'.

5.1.1 Current Mobility Behaviour

The panel member's current mobility behaviour is characterised by a strong manifestation in car ownership. 54.4%[2] stated to possess a car whereas 22.1% are in possession of more than one vehicle. Solely 11.2% reported to not owning a private car. The differences in car ownership among the analysed generations should be seen as insignificant. In accordance with previous publications (cf. e.g. Henkel et al., 2015, 3 f.) the security of a specific level of individual mobility or rather personal freedom is seen as the main rationale for possessing a private vehicle (73.5%). Nevertheless, nearly half of the participants indicate that in their view a car merely serves the purpose as a means to an end to overcome spatial distances (49.5%). Participants who indicate an emotional bond with their vehicle solely represent 26% of the panel members. This could indicate a confirmation of the theory of 'de-emotionalisation' of cars as discussed amongst others by Bratzel (2014, p. 95). The follow-up question regarding the main rationale for not possessing a private vehicle could provide further evidence for this assumption. A qualitative categorisation of the open question responses of participants that do not possess a car indicates that the availability of sufficient transport alternatives (e.g. public transport) in combination with the high costs associated with cars are the main rationales advocating against ownership.

Concerning the usage of information in the travel mode selection, it could be claimed that in most cases a comparison is solely a reaction to extraordinary events, which might suggest a strong habituation of those processes as advocated by Lenz (2011, p. 613). 54.2% compare alternative modes of transport if concrete information on feasible events that could have a negative impact on the travel time such as a high likeliness of congestion or weather forecasts is at hand. In addition, 43% indicate a comparison if the destination or rather the most appropriate route to the destination is unknown. 45.2% report the comparison of transport modes if sufficient time before departing is available. Real-time information such as push notifications concerning current congestion levels are used as a trigger for comparison by 12.1% whereas 16.4% never compare travel mode alternatives. The most frequently used devices to gather and compare travel-related information are smartphones (80.6%), followed by home computers/laptops (77.4%) and tablets (39.1%).

---

[2]Please note that the percentage in the following paragraphs is calculated on the basis of total number of interviews not on the total number of responses to the respective question.

The qualitative categorisation of open question responses of participants that reported no comparison of transport mode alternatives complies with findings of previous studies discussed amongst others by Delatte et al. (2014, p. 9). The most frequently stated reason is the perception of unattractiveness of available alternatives, mostly in terms of flexibility. Moreover, the high effort for comparing alternatives as well as the habitual selection of different transport modes for different occasions or routes was regularly stated.

Mobility services are less regularly used by the survey panel. 41.3% do not use any of the presented mobility services while 42.9% use car rental services occasionally. Car sharing and ridesharing services are utilised by 15.4% and 9.9%, respectively. As main rationales for the low usage insufficient service coverage as well as the inflexibility of usage are stated.

5.1.2 Customer-Centric Management in Mobility Services

The utilisation of CRM measures in terms of loyalty programmes is widely accepted by the survey's participants. Merely 19.9% perceive customer loyalty schemes for mobility services as unattractive. Nonetheless, the most selected rewards and amenities are rather traditional options such as price reductions for future mobility service purchases (37.8%) as well as the exchange of accumulated 'mobility kilometres' for certain benefits (37%). A rather novel amenity that was repeatedly selected is the feasibility to use dedicated traffic lanes similar to high-occupancy lanes in the United States (32.5%). This high attractiveness might be traced back to the high ratio of car owners in the survey panel, however.

The integration of VAS and the attractiveness of using those services varies vastly among the presented options as well as among the generations. From an age-independent point of view services which are not primarily concerned with the provision of mobility services received fewer positive receptions. As least attractive or rather services which are less likely to be used are NBO services, where 67.7%[3] demonstrated disinterest in utilisation. With the exception of universal design, which received the highest likeliness of usage (69.1%[4]), and the availability of different SLAs (50.6%) all other presented service categories received a likeliness of usage below 50%.

A deeper analysis on the basis of generation affiliation reveals that 'Baby Boomers' are more conservative in their willingness to VAS usage, specifically in comparison to members of 'Generation Y'. The largest discrepancy can be observed in entertainment services. While 67.5% of 'Generation Y' participants express their

---

[3]The unattractiveness of services is based on the sum of respondents selected options stating that a usage is fairly unlikely and certainly excluded.

[4]The attractiveness of services is based on the sum of respondents selected options stating that a usage is rather likely and certain.

willingness in the utilisation of such services, merely 28.3% of 'Baby Boomers' express their willingness in the same manner. Similar results can be observed in the usage of personalisation options. More than half of the participants (52.2%) affiliated with 'Generation Y' are interested in the personalisation of services. On the other hand, solely 31.1% of 'Baby Boomers' and 33.5% of 'Generation X' find personalisation options attractive. The least attractive service category according to 'Generation Y' is the availability to select from different SLAs where merely 46.5% express their interest in utilisation ('Baby Boomers': 50%; 'Generation X': 54.3%).

As the largest barrier towards customer-centric mobility the lack of sufficient service coverage (45.2%) was stated. Moreover, insufficient legal protection in terms of adaptation of existing laws and ordinances (44.7%) was of great concern to the participants. The strong car-centred culture as well as the strong car lobby in Germany is the third most selected barrier (40.8%).

5.1.3 Asset-Driven Versus Data-Driven Companies

The subconscious survey in terms of capital investment demonstrates a clear tendency towards digital and non-traditional mobility companies. The largest share of participants is interested in investments in Google (42.8%) which received slightly more attention than Tesla (42.1%). Apple is ranked third with 28.1%. The first 'established' mobility provider is BMW which is ranked fourth with 26.8%. A noticeable common characteristic of those companies is their involvement in the development of smart cars and ADV. The least selected companies were third-party railway companies (2.9%) and the ridesharing platform BlaBlaCar (3.9%).

The most selected business model might advocate for a preference of data-driven companies as well. 43.9% favour the aggregational approach. The integrational and the hybrid approach as described by van Audenhove et al. (2014, p. 23) are both favoured by 23.9% of the survey panel. The business model of a 'digital transport association' received 3.7%. 4.6% of the participants prefer a non-digital business model. A split by generation does not indicate significant discrepancies with exception of the aggregational approach. Close to half of the participants belonging to 'Generation Y' (49.7%) selected this business model, which is the strongest manifestation in item selection within the cross-generational analysis. Furthermore, the 'digital transport association' model as well as the non-digital model is more preferred by participants of 'Baby Boomers' and 'Generation X'. For instance, 52.4% of the participants that selected a non-digital business model belong to 'Generation X' (in comparison to 14.3% of 'Generation Y').

Digital companies also receive the most positive reception concerning the execution of several tasks associated with their respective business models. The competence of the provision of comprehensive VAS offerings is clearly attributed to internet companies (50.9%) and technology firms (14.7%). The ability to realise a seamless integration of traffic carriers (integrational approach) is awarded to Internet companies (23.0%), transport associations (21.1%) and railway companies (17.5%). The operation of a business model following an aggregational approach is seen in the field of competence of Internet companies as well (37.7%). The establishment of a hybrid business model is again awarded to Internet companies (24.1%). Nevertheless, the gap between data- and asset-driven companies in this category could be seen as insignificant as car manufacturers rank second with a marginal difference in frequency (23.9%). The sole category where the lead is seen in asset-driven companies is the operation of a business that supplies cities or regions with complete mobility systems. In this category transport associations are distinctively perceived as the most competent by nearly half of the participants (46.5%). Notwithstanding, Internet companies are ranked second (13.4%) ahead of public institutions (11.2%).

5.1.4 Future Requirements for Mobility Services and Providers

The utilisation of ADV in future mobility systems divides the survey panel. 55.9% perceive smart cars as attractive or rather very attractive, whereas 24.3% completely reject such vehicles. Situations in which ADV are perceived as most useful are situations in which the alleviation of mental and physical effort can be achieved. 58.8% see the greatest usage potential on long-distance trips as well as to increase road safety at certain locations (43%) and temporary automation in order to avoid or dissolve congestion (38.8%). 14.9% indicate no usage of ADV at any time. These findings could argue for the cultural barrier mentioned by Kröger (2015, p. 64). It might be claimed that ADV are generally regarded as driver support rather than an actual replacement of the driver itself. Merely 22.6% of the participants welcome a comprehensive and ubiquitous usage of ADV.

The evaluation of attributes (future) mobility services have to possess could imply that assumptions made in Sect. 3.1 concerning the alteration of buying behaviour can be supported by empirical data gathered in this research. Convenience and comfort are seen as most important requirements participants place on mobility services which could be seen as confirmation of suggestions made by Winterhoff et al. (2015) who claim that customer experience might be the strongest decision factor in mobility services. 92.75% indicate that reaching a destination in a convenient and time-efficient manner is more important than comparing routes or operators. In addition, 85.27% place a higher importance on comfort and convenience than on the brand of a mobility service provider. For 52.7% comfort is even of greater importance than price. This complies with the findings of Accenture (2014) in other industries and could indicate a further commoditisation of mobility services and the erosion of customer loyalty. This may be supported by the strong agreement on the unimportance of a mobility service provider's brand in the operator selection process as well. Furthermore, 83.96% regard a consistent service level throughout mobility chains as important which might also be seen as related to the importance of comfort. The experience of mobility service providers is considered by more than half of the participants. 51.65% indicate that they rather trust established providers when selecting an operator. Besides, 83.30% of the survey panel reported that mobility should be integrated into their everyday life to a higher degree.

5.2   Validation of Results

As advocated by Skowronek and Duerr (2009) as well as Callegaro et al. (2015, 54 f.), the verification of data drawn from non-probabilistic samples with methods derived from inferential statistics should be regarded as a valuable contribution, despite the potential for actual inference. Moreover, in practice, inferential statistics are regularly applied to non-probabilistic data (cf. Callegaro et al., 2015, p. 54).

The required sample size for a confidence level of 95% as well as a margin of error of 5% was calculated using the following formula:

$$\text{Sample Size} = \frac{\frac{z^2 * p(1-p)}{e^2}}{1 + \left(\frac{z^2 * p(1-p)}{e^2 N}\right)}$$

N = the total population; e = margin of error; z = z-value

As the total population size, the population of Germany excluding inhabitants under 6 years was assumed which equals 77.64 mn[5] (=78 mn for the calculation). As z-value 1.96 was determined. Assuming a normal distribution (50%) a sample size of 385 would be required. With an actual sample size of 456 a margin of error (ME) of 4.59% can be achieved. Nonetheless, as highlighted by Callegaro et al. (2015, p. 54), since the data was gathered with non-probabilistic methods the true ME cannot be calculated properly and therefore might exceed the calculated value.

5.3   Discussion of Findings

The survey's findings could be regarded as a confirmation of the anecdotal image of Germany as a car-centred culture. Many participants indicated that their individual level of mobility could solely be maintained by owning a private car. Nevertheless, a trend towards the rationalisation of a private car's status can clearly be observed as well. It may be stated that the emotional bond between a car and its owner does not necessarily impact the frequency of usage or rather the comparison of transport mode alternatives. Out of the participants that indicated no comparison of transport modes persons who reported an emotional bond with their car and persons that see cars as a means to an end are almost equally distributed with 37.5% and 38.9%, respectively. In addition, the number of children does not have any statically significant impact on car ownership in this sample (Eta = 0.066). Therefore, the strong importance of convenience might be seen as an overarching determinant in this regard.

Further, the findings contain noticeable evidence for the application of subjective filters in the transport mode selection process. As mentioned by Bartz (2015, p. 42) as well as Delatte et al. (2014, p. 9) alternatives to private cars are perceived as

---

[5] Source: Destatis (2016).

generally unattractive by participants which do not compare alternatives. Moreover, the high effort of gathering and comparing information is seen as a large barrier, which advocates for the vast demand for convenience as well. This may indirectly support the findings of Grotenhuis et al. (2007, 34 f.) who argue that information is solely valuable if they enable time and effort savings as well as the suggestions of Otto (2010, p. 2) who sees information as a key requirement for the abolishment of habits in the transport selection process.

The survey participants' attitude towards CCM practices in mobility services may confirm further assumptions made in previous chapters. The vast discrepancy among the analysed generations may be seen as a confirmation of the requirement for truly personalised customer experiences (i.e. CEM). As mentioned before, younger generations seem to perceive VAS which are not directly related to the provision of transport services more valuable than previous generations. The especially strong diversion in attitude towards entertainment and NBO services could be primarily related to the technological environment a research participant was raised in. Members of 'Generation Y' are often referred to as 'digital natives' which should indicate their strong affiliation to digital technologies such as computers and the Internet. Thus, the attitude towards these types of services could be perceived as more natural or intuitive as by previous generations (so-called 'digital immigrants').

A correlation analysis reveals moderate, statistically significant correlations between the age and the willingness to utilise non-mobility services, which could be seen as slight confirmation of the previously made assumptions. Accordingly, the strongest correlation is found in entertainment services ($r_s = -0.332$; significance level $= 0.01$), followed by location-based NBO ($r_s = -0.231$; significance level $= 0.01$). The enclosure of places of interest does not demonstrate as strong correlations as NBO services ($r_s = -0.118$; significance level $= 0.05$). This development could be regarded as trend, nevertheless. Similar to the trend of declining car ownership the acceptance of non-mobility services may increase over time. Moreover, as no responses of 'Generation Z' were obtained the attitude of affiliates of this generation is unknown and neglected in the results of this research. Results of the pretest to this research, where a distinctively younger panel was generated, suggest that the acceptance of non-mobility services could be considered high in this generation, however. Thus, the feasibility of individualising mobility services could not only be understood as enabled by digitalisation, but also as demanded on the customer side. This might highlight the importance of true CCM once more.

Another factor that might influence the willingness to usage of certain VAS is the concern over (data) security. According to Hofstede's cultural dimensions model, Germany is positioned among countries with a relatively high level of uncertainty avoidance (cf. Hofstede Centre, 2016). The absence of sufficient legal protection in terms of data security and the safeness in the usage of technological innovations (e.g. ADV) might influence the attractiveness of certain VAS. An investigation for empirical proof in terms of a correlation analysis between the selection of legal uncertainty and data security as well as the attitude towards the utilisation of car boot deliveries and NBO services does not demonstrate considerable interrelations, however. Notwithstanding, it may be argued that such interrelations cannot be explained to a satisfactory level with the available variables in the research.

In contrast to the conservative attitude towards VAS, the selection of mobility service providers could be considered less traditional. Yet, this may be regarded as a confirmation of the strong and intensifying commoditisation of the industry as well. Statements with respect to the requirements on mobility service indicate that the service itself is of greater importance than the company providing it. Comfort and convenience might be regarded as the most important decision criteria. This assumption might be encouraged by the most frequently selected business model. Since nearly 44% of the panel indicated to regard the aggregational approach as more attractive, it could be assumed that the feasibility to compare several mobility services providers appears to be important to the participants. Notwithstanding, current practices in the comparison of transport modes suggest that a convenient manner of information gathering has to be regarded as prerequisite. On the other hand, it should be considered that the integrational as well as the hybrid approach are both selected by 23.9% of the participants which equals 47.8% of the panel. This may demonstrate the trade-off between price and convenience a customer is confronted with. It thus might be argued that more price-sensitive customers are more likely to select the aggregational approach whereas customers who appear to be more convenience-oriented are more likely to select the integrational or hybrid approach. Since a participant's income was not collected in the survey this assumption cannot be confirmed with the available data, however.

The relatively low importance of a mobility service provider's brand may have implications on suggestions presented by Winterhoff et al. (2015, p. 23). The authors suggest that due to the abundant availability of AMODS customers may reduce the number of private vehicles in their households and therefore trade up to premium brands. Since this research's panel does not indicate brand image as a strong decision factor this assumption might be evaluated as critical. Nonetheless, the distinction between a mobility service provider and automotive OEM may not be as clearly pointed out as required to draw a reliable conclusion or implication. Besides, it might be argued that the maybe still a market for premium services and vehicles in future mobility markets. Nevertheless, the unimportance of brand as well as the relatively large share of participants that do not solely rely on established mobility service providers could pave the way for new market entrants of digital origin.

References

Accenture. (2014). *Customer 2020: Are you future-ready or reliving the past?: Ten years of Accenture research highlights real opportunities for providers to better meet customers' steadily rising expectations* (No. 14-6613). Accenture.

Bartz, F. M. (2015). Mobilitätsbedürfnisse und ihre Satisfaktoren. *Die Analyse von Mobilitätstypen im Rahmen eines internationalen Segmentierungsmodells.* (Doctoral Thesis). University of Cologne, Cologne.

Bratzel, S. (2014). Die junge Generation und das Automobil – Neue Kundenanforderungen an das Auto der Zukunft? In B. Ebel & M. Hofer (Eds.), *Automotive Management. Strategie und Marketing in der Automobilwirtschaft* (2nd ed., pp. 93–108). Springer.

Callegaro, M., Manfreda, K. L., & Vehovar, V. (2015). *Web survey methodology.* Sage.

Delatte, A., Kettner, S., Schenk, E., & Schuppan, J. (2014). *Multimodale Mobilität ohne eigenes Auto im urbanen Raum: Eine qualitative Studie in Berlin Prenzlauer Berg.* Technical University of Berlin.

Destatis. (2016). *Bevölkerung.* Retrieved August 10, 2021, from https://www.destatis.de/DE/ZahlenFakten/GesellschaftStaat/Bevoelkerung/Bevoelkerung.html

Grotenhuis, J.-W., Wiegmans, B. W., & Rietveld, P. (2007). The desired quality of integrated multimodal travel information in public transport: Customer needs for time and effort savings. *Transport Policy, 14*(1), 27–38. https://doi.org/10.1016/j.tranpol.2006.07.001

Henkel, S., Tomczak, T., Henkel, S., & Hauner, C. (2015). *Mobilität aus Kundensicht.* Springer Fachmedien Wiesbaden.

Hofstede Centre. (2016). *What about Germany?* Retrieved August 10, 2021, from https://geert-hofstede.com/germany.html

Kröger, F. (2015). Das automatisierte Fahren im gesellschaftsgeschichtlichen und kulturwissenschaftlichen Kontext. In M. Maurer (Ed.), *Autonomes Fahren. Technische, rechtliche und gesellschaftliche Aspekte* (pp. 41–67). Springer.

Lenz, B. (2011). Verkehrsrelevante Wechselwirkungen zwischen Mobilitätsverhalten und Nutzung von IuK-Technologien. *Informationen zur Raumentwicklung, 10,* 609–618.

Nicholas, A. J. (2009). Generational perceptions: Workers and consumers. *Journal of Business & Economics Research, 7*(10), 47–52.

Otto, S. (2010). *The psychology of transport choice.* Institute for Ecological Economic Research (IÖW).

Skowronek, D., & Duerr, L. (2009). The convenience of nonprobability: Survey strategies for small academic libraries. *College & Research Libraries News, 70*(7), 412–415.

van Audenhove, F.-J., Korniichuk, O., Dauby, L., & Pourbaix, J. (2014). *The future of urban mobility 2.0: Imperatives to shape extended mobility ecosystems of tomorrow.* Retrieved from Arthur D. Little/The International Association of Public Transport (UITP) website: www.adl.com/FUM2.0

Winterhoff, M., Shirokinskiy, K., Mishoulam, D., Freitas, N., & Chivukula, V. (2015). *THINK ACT automotive 4.0: A disruption and new reality in the us?*

Scenario and Model Development

6

6.1 Future Scenarios of a Digitalised Mobility Market

Initially applied solely in the military sector, the development and management of scenarios has been introduced in strategic management during the 1950s and 1960s in order to enhance a company's long-term planning (cf. Meinert, 2014, p. 8). Scenarios are utilised to construct several feasible alternative futures in order to derive key issues, vulnerabilities as well as possibilities for adaptation measures (cf. Mahmoud et al., 2009, p. 799). According to Schoemaker (1995, p. 27) scenarios should be seen as mitigation between the over-prediction and under-prediction of change in long-term planning. Nevertheless, scenarios should not be regarded as predictions or forecasts, but rather explorations of feasible futures derived from current developments based on causalities (cf. Meinert, 2014, p. 8). The scenario development in this research followed the framework provided by Meinert (2014, 13ff.). In a first step, the time horizon was determined. Since the underlying issue of the scenario development equals the central research problem of this research the problem definition was regarded as obsolete. The time horizon or rather the year the market situation should be explored was determined to 2030. The main rationale behind this is that some key elements of digital mobility systems such as ADV are still in the development phase. Moreover, it was assumed that after the introduction of these new mobility system components a certain time is required to establish such novelties in the market. The adaptation of laws and ordinances may require a certain time as well.

After the time horizon determination key drivers of change were identified. The identification processes considered both macro-environmental developments derived from the literature review as well as key findings of the questionnaire survey which were considered as quantification of specific mobility trends. Notwithstanding, a qualitative review was applied since the survey results may have been affected by an uninformed response bias (URB) on the part of the survey panel. URB describes the willingness of participants to express their opinion on issues in which their knowledge and level of information is insufficient or a participant is

P. Siegfried, *Digitalisation in Mobility Service Industry*, Future of Business and Finance, https://doi.org/10.1007/978-3-031-07151-5_6

completely unfamiliar with (cf. Graeff, 1999, p. 632). Even though this bias was mitigated by providing a participant with the option to waive certain questions, in some areas of the questionnaire this was seen as inapplicable.

One example of URB may be the selection of a suitable company for the respective business models. For the aggregation approach transport associations received the strongest positive perception after Internet companies. Nevertheless, the complete transformation from an asset- to data-driven company could be regarded as unlikely. It could, however, be assumed that participants confused the business model of an integrator and aggregator since some platforms of integrators (e.g. RMV, Quixxit and Moovel) allow a comparison of routes and transport mode alternatives. Therefore, it was decided to assess the survey results (i.e. the quantitative driver evaluation) qualitatively to a varying degree in the scenario development process.

After the major drivers and 'givens' were identified two scenarios were developed and described in a narrative manner. The main driving factor separating both scenarios is the degree of influence digital companies might gain in future mobility markets. The first scenario (see Sect. 6.1.1) describes a digital mobility market as quantitatively as possible based on the survey's results. The influence of digital companies hence is relatively high. The second scenario (see. Sect. 6.1.2) on the other hand considers the previously mentioned qualitative interpretation to a larger extent. The influence of digital companies is considered as moderate to high. The last step mentioned by Meinert (2014) in terms of a scenario reflexion is encompassed in chapter seven.

Both scenarios follow several overarching assumptions which should be stated in order to enhance the comprehensibility of the following paragraphs. First of all, it was implied that in 2030 full legal certainty is given. This includes legal protection with regard to data security as well as amendments to current laws and ordinances concerning liability in cases of accidents with unmanned vehicles. Moreover, a complete liberalisation of the German mobility market was assumed. This comprises the regulation on taxi services as well. Recent support claims of the European Union (cf. Fioretti, 2016) concerning the introduction of services provided by firms such as Uber and Airbnb were seen as an indication for this trend. Furthermore, sufficient coverage of both mobility services and (mobile) Internet is given.

6.1.1   The Data Revolution

In 2018, Google launched its service 'Google Go' to design individual mobility chains based on the access of several available mobility service providers, which has evolved into the market-leading product by 2030. Other Internet companies fast followed 2 years after Google's market entry which led to a strong competition between established mobility service providers and the new entrants. In cooperation with volume car manufacturers, Amazon commenced AMODS in 2025, initially for its 'prime' members, to combine online shopping and the entailed deliveries with personal mobility. One year later, these services were expanded by the integration of further mobility services such as bike sharing and made available to the whole customer base of Amazon. Google's long-awaited ADV which was developed in cooperation with Fiat has been introduced in the course of the commencement of Fiat's AMOD service line 'mobilità'. Continuing its close co-operation 'mobilità' was initially made available exclusively on 'Google Go'. Facebook pursued a similar approach. By forming strategic alliances with companies such as Tesla and third-party bike sharing firms, Facebook began offering mobility services adapted to its customers' needs. Even though Facebook relied upon the integration of offered service components to a certain extent, for instance, with the commencement of autonomous driving long-distance bus services, the company rather focused on the formation of strategic partnerships with firms that complement Facebook's own services.

After 5 years of harsh competition in the battle for supremacy, many established mobility service providers shifted their focus to niche markets and other mobility-related activities. Due to the strong competition in the volume segment car manufacturers such as Daimler and BMW focussed on the premium segment since Amazon and Facebook are not only able to realise cheaper fares, but also to customise mobility services more accurately. The volume segment is clearly driven by the aspect of price and highly commoditised. Despite the strong performance and reputation in the premium segment, Daimler and BMW are strongly dependent on aggregating platforms such as 'Google Go' in order to achieve an economically viable asset utilisation in the volume market. As a consequence, BMW recently announced to phase out and sell its AMOD service line 'DriveNow'.

The premium segment is suffering from commoditisation as well. As customers do not regard brand as a strong decision criterion, Amazon is preparing the launch of a premium line to gain market share in this segment. Transport associations (e.g. RMV) concentrate on rather rural areas. The increasing urbanisation has led to unattractiveness of non-urban regions. Since for historical reasons transport associations possess the densest network in such areas, it was decided to focus on these markets. Although transport associations are present in urban regions as well, most traffic is generated through platforms such as 'Google Go' as transport associations have failed to adapt to the customers' demand for customisation as quickly as required. Thus, such companies often serve as mere transport and infrastructure providers as well as mobility building blocks for customers of Google and Facebook.

Due to stricter regulations regarding the pursuit of a more sustainable development, the German government has passed several laws which put HMP responsible for the management of sustainable urban transport systems. Therefore, HMP use previously introduced customer loyalty programmes to provide incentives to switch to other transport modes or departure times as well. The majority of these loyalty schemes rely upon the accumulation of 'mobility kilometres' which are dynamically calculated and dependent on current congestion levels, occupancy levels etc. Fines HMP face for surpassing certain measurement thresholds ('Sustainability KPIs') are directly taken into consideration in the price calculation. Hence, customers that behave more sustainably (e.g. by increasing vehicle occupancy) will receive more 'mobility kilometres' and vice versa.

In 2030, traffic still consists of ADV and non-ADV. The occasions ADV take over the full control over itself are temporarily constrained. Most ADV solely provide valet-parking services. In order to enhance the traffic flow in urban areas, dedicated ADV lanes were established. In addition, in high congestion occasions or the anticipation of high congestion levels ADV take over the control automatically in order to stabilise and maintain a city's traffic flow.

6.1.2 Competition 4.0

After steadily increasing two-digit growth rates of mobile mobility applications such as 'Quixxit' and 'Moovel' the Association of German Transport Companies (VDV) initiated a project in 2017 which aimed for the establishment of an own mobility application to increase the competitiveness of public transport as well as profit from this new market trend. The mobility management platform 'Ubiquity' was launched in 2020 which provided end customers with the feasibility to travel seamlessly across Germany. In order to achieve this, the VDV and its affiliates had to undergo substantial changes: Besides the integration and harmonisation of IT and billing systems, all transport means were equipped with free Internet access and power sources as well as (at least partially) automated. Moreover, advanced forecasting and right time analytics systems had been introduced to ensure seamless travel. The fare schemes were modernised as well. Passengers are billed after a trip is performed according to their actual usage of transport means. In rural regions, the VDV acts as a full-service operator by supplying municipalities with complete mobility systems.

Google's competing product 'Google Go' was launched in 2020 as well. Especially third-party mobility service companies and volume car manufacturers derived benefits of Google's market entry as 'Google Go' provided a platform to generate additional traffic. In contrast to VDV's 'Ubiquity' which intends to integrate mobility into everyday activities, Google's unique selling proposition (USP) is the fusion of mobility and everyday tasks to respond to the increasing demand for 'hyper convenience'. By heavily relying on NBO and entertainment services 'Google Go' is specifically targeting younger generations.

Premium car manufacturers such as Daimler and BMW massively expanded their service portfolio and coverage after the market entry of Google in order to remain competitive. After the introduction of ADV in 2025 Daimler and BMW began to utilise this new technology as their 'backbone service component' and shifted their focus to individual mobility (i.e. 'new motorised individual transport'). Due to their historical origin, both companies see their USP in the integration of private vehicles in the overall mobility chain. Based on the customers' individual needs comprehensive service packages are offered which also include entertainment and e-commerce services realised through co-operations with Netflix and Spotify. In addition, Daimler and BMW mobility platforms are connected to public transport operators. Nevertheless, the firms' service philosophy aims for the provision of own services to the highest degree possible. Public transport alternatives are mostly offered if parking space forecasts or AMODS utilisation levels are suggesting a low probability of availability.

Similar to scenario one ADV users are free to select between driving and being chauffeured on dedicated traffic lanes. Nonetheless, specifically urban centres have been declared mandatory automated driving zones since these areas are prone to congestion and demonstrate a high likeliness of accidents. By entering such zones ADVs automatically take over the control of itself.

Between 2020 and 2030 car ownership declined drastically. The most significant decline was observed in 2027 when AMODS eventually became available across Germany. As a result, the car volume segment is almost completely substituted by AMODS in 2030. Even though the premium segment is still dominated by car ownership, a declining trend is currently observed as well. This trend is regarded as an opportunity by Amazon which recently announced the launch of a premium service line in close co-operation with Tesla.

The increase in competition and the abundance of information available online has led the mobility service industry into rigorous price wars. Extensive CRM programmes have been introduced in order to retain customers. Volume segment operators such as Google and Facebook mainly rely upon the principle of accumulating 'mobility kilometres' and their exchange for rewards and amenities such as price reductions, free access to entertainment platforms or service upgrades. Premium service operators such as Daimler and BMW implemented loyalty schemes, which follow the example of airlines. Frequent users receive specific statuses which increase with the service usage. Depending on the respective status users are granted preferential booking as well as the guaranteed availability of AMODS.

6.2 The Digital Mobility Split

After a thorough discussion on the impact digitalisation might have on the mobility service industry it appears crucial to assess these effects quantitatively. As mentioned earlier, the traffic volume of a city or region is traditionally measured with the calculation of the modal split. Nevertheless, the utilisation of this instrument might not be appropriate as its calculation is disregarding two factors that will increase even further in course of the digitalisation. One factor which is currently unembodied is the increasing share of multimodal and intermodal travel. Since the modal split categorises traffic volume according to the available modes of transport multimodality is not effectively portrayed. The second factor might be related to the established categorisation scheme as well. It may be argued that the modal split describes traffic volumes from a rather asset-based view. Since it is advocated by several studies that the majority of mobility services or rather future mobility will be intermodal or multimodal this asset-based perspective may become obsolete. Thus, in course of the industry's transformation, the manner in which the traffic output will be measured should be altered as well.

It is, therefore, proposed to measure the traffic volume of digital mobility systems from a service-based view. The main rationale should be seen in the rising importance of services and servitisation. The following paragraphs ought to construct a novel model for such measurement, which hereafter will be referred to as digital mobility split (DMS).

6.2.1 Method of Calculation

Similar to the modal split the DMS should quantify and depict the proportion of usage of the available mobility options based on a predetermined parameter. In order to encompass the previously mentioned factors of multimodal mobility and servitisation it was decided to calculate the DMS on the share of usage of an operator or rather the proportion of service bookings attributed to the respective HMPs.

Due to the inexistence of the discussed HMP business models at present state the calculation has to include several assumptions. In order to estimate the future usage of an operator, it was implied that the selection strictly follows a participant's personal preferences on mobility services as collected in question 17 of the questionnaire. The sole consideration of the preferred business model was considered insufficient, mainly due to the previously mentioned URB. As survey participants are currently unable to actually experience holistic mobility services an individual might not know which business model is the most suitable according to their preferences.

It was assumed that the preference for a service provider is dependent on three factors: the importance of (consistent) service levels, the importance of a service provider's brand as well as the industry reputation of a company (i.e. the degree of reliance on traditional service providers). To a certain extent, these factors had been implemented in the design of Question 17. In order to identify variables that represent these factors all variables encompassed in Question 17 were analysed on their response distribution. It was discovered that out of the seven variables four demonstrated high mean values. Since this may bias the calculation, it was decided to develop the user preference model on the three remaining variables.[1]

After the user preference model was developed a scheme for the allocation of preference profiles to the respective business models was created. The allocation scheme had to follow several assumptions as well. It was decided that the DMS should solely possess four instead of five categories. This has two reasons. Firstly, if a region decides upon the provision of mobility services through a 'digital transport association' it would automatically exclude other service providers. Secondly, this business model could be regarded as a special form of the integrational approach which might allow the creation of a hybrid approach as well. Thus, to enhance the

---

[1] All variables with a mean score below 3 have been considered in the preference model computation.

Table 6.1 Preference profile archetypes and their most suitable business model

Preference profile	Business model
• Places importance on brand • Trusts in rather traditional mobility service provider • Places high importance on service	Integrational Approach
• Places little to no importance on brand • Trusts new mobility service operators as well • Places less to medium importance on service	Aggregational Approach
• Places little to no importance on brand • Trusts new mobility service operators as well • Places medium importance on service	Hybrid Approach

Source: Own depiction

mutual exclusiveness of the DMS's categorisation it was renounced to establish an additional category for this business model. The determination of the model categories was followed by the formulation of archetype preference profiles attributed to every business model. The resulting allocation scheme is summarised in Table 6.1.

The actual DMS was achieved by performing a k-means clustering analysis. As a first step participants who indicated a preference for non-digital business models were removed from the data to be analysed. Therefore, number of clusters was predetermined to three as non-digital users should be considered as-is in the DMS. After the clustering process, the created clusters were compared to the archetype preference profiles presented in Table 6.1 and assigned based on the lowest diversion from the respective archetype.

6.2.2 The Digital Mobility Split Applied to the Sample Data

The application of the DMS on the sample data demonstrates noticeable similarities to the question surveying the most preferred business model. According to the DMS, the majority of participants would prefer the usage of services provided by aggregators. The usage of genuine integrational services is less frequently observed than in question 14 of the questionnaire, however. Consequently, the usage of hybrid services demonstrates the second largest category. The final split is summarised in Table 6.2.

Since findings thus far indicated a strong deviation in opinions between the represented generations the DMS should be analysed in this respect as well. The majority of participants affiliated to the generation of 'Baby Boomers' demonstrate a user preference profile associated with the hybrid approach (39.1%). In contrast, usage profiles of participants belonging to 'Generation X' and 'Generation Y' are rather associated with the aggregational approach (41.8% and 43.6%, respectively). While members of 'Generation Y' show equal tendencies towards a hybrid and

Table 6.2 Digital mobility split applied to the sample data

Category (Cluster number)	Number in cluster	Percentage
'Digital Neanderthal men' (0)	22	4.48
Hybrid Approach (1)	141	30.99
Aggregational Approach (2)	182	40.00
Integrational Approach (3)	111	24.40

Source: Own depiction

integrational approach, affiliates of 'Generation X' show stronger preferences for a hybrid model. This could suggest that preference profile change towards a higher importance of service as the age of a participant increases.

References

Fioretti, J. (2016). *EU throws support behind 'sharing economy' firms like Uber, Airbnb.* Retrieved August 16, 2021, from http://www.reuters.com/article/us-eu-services-idUSKCN0YO12I

Graeff, T. (1999). Uninformed response bias in measuring consumers' brand attitudes. *Advances in Consumer Research, 26,* 632–639.

Mahmoud, M., Liu, Y., Hartmann, H., Stewart, S., Wagener, T., Semmens, D., & Winter, L. (2009). A formal framework for scenario development in support of environmental decision-making. *Environmental Modelling & Software, 24*(7), 798–808. https://doi.org/10.1016/j.envsoft.2008.11.010

Meinert, S. (2014). *Leitfaden Szenarioentwicklung.* European Trade Union Institute (ETUI).

Schoemaker, P. J. H. (1995). Scenario planning: A tool for strategic thinking. *Sloan Management Review, 36*(2), 25–40.

Conclusions for the Digitalisation in Mobility Service Industry

7.1 Conclusions

Digitalisation has been attributed to the potential to alter the mobility service industry completely. Reduced market entry barriers and new possibilities to provide services in a novel and convenient manner are regarded as key indications for a paradigm shift in the industry. First signs of such a digital disruption are already observed in terms of increasing availability of mobility-related services provided by internet companies and technology firms. The involvement of those companies in the development of ADV is evaluated as an indication for a potential shift of paradigm as well. The aim of this research was to explore and assess the potential implications digitalisation may have on mobility services and mobility as a whole. This included the analysis of current mobility behaviours, the likeliness of a paradigm shift from plain mobility management to CCM as well as a shift in the market dominating companies. In addition, requirements that could be placed on future mobility services and its providers should be derived. After a thorough analysis, the research questions stated in Sect. 4.1 should thus be answered.

The present mobility behaviour of the survey panel is strongly characterised by MPT. The comparison of transport mode alternative is provoked by extraordinary events such as new destinations or the occurrence of events which might impact the travel time negatively. Moreover, the comparison might be regarded as strongly biased by subjective filters applied to available information. Mobility services available in their current state are merely used by a minority of the survey participants, mostly due to a lack of service provision as well as the lack of flexibility and administrative hurdles. The selection of potential mobility service providers is guided by the factors of comfort and convenience. For the majority of the survey panel, these factors are of greater importance than an operator's brand. More than half of the participants' value comfort even higher than price.

The implementation of CRM or rather CCM activities in mobility services such as loyalty programmes is widely accepted. Nevertheless, the most preferred rewards are rather traditional comprising price reductions for future purchases and service

© The Author(s), under exclusive license to Springer Nature Switzerland AG 2022
P. Siegfried, *Digitalisation in Mobility Service Industry*, Future of Business and Finance, https://doi.org/10.1007/978-3-031-07151-5_7

upgrades, amongst others. In addition, VAS which are not primarily related to the alleviation of overcoming of spatial distances are perceived as less attractive, even though the integration of mobility into everyday life activities is strongly desired by the survey panel. This might highlight the function of mobility which can be derived from its definition. Mobility is mainly perceived as a secondary need. Hence, it may be argued that mobility should be integrated into a customer's life for the purpose to lessen the effort required for the selection of the most appropriate transport mean. A fusion of mobility and other services of customer interests may be regarded as a too strong interference in the individual's privacy or freedom of choice. Factors influencing the attractiveness of certain VAS could not be explained with statistically significant evidence. Notwithstanding, the findings suggest a trend towards the acceptance of non-mobility-related VAS in younger generations. Thus, it could be claimed that the importance of these services will increase over time. This may be attributed to the technological environment an individual was raised in. Younger generations such as the 'Generation Y' and the 'Generation X' may demonstrate a stronger affiliation towards location-based NBO and entertainment services as the usage of these types of services could be perceived as more natural and intuitive by older generations. The largest barriers to customer-centric mobility services are seen in the lack of comprehensive service coverage, legal uncertainty as well as the strongly car-centred culture in Germany.

The findings further suggest that the way for the market entry of firms of digital origin is paved. Companies associated with digital products and services are the preferred investment object. Specifically, companies that are involved in the development of ADV seem to be the most attractive. Moreover, the aggregational business model is the single most selected item which is clearly attributed to companies of digital origin (cf. Rammler & Sauter-Servaes, 2013, p. 45). Nevertheless, a cumulative percentage of 47.8% see the most preferable business model in the integrational approach and a hybrid version of both business models. Except for the business model of a full-service operator, which provides complete mobility systems, Internet firms such as Google or Facebook are regarded as the most competent in the fulfilment of tasks associated with the respective business model as well as the business models itself. Furthermore, the industry reputation of a provider is merely considered by circa 50% of the panel. This could be seen as additional factor favouring the market entry of data-driven companies. The application of the DMS developed in this research verifies the trend towards digital companies or rather the aggregational approach. Therefore, it might be concluded that from insights founded on the empirical data gathered in this research, digital companies are currently perceived as the most competent to inherit the role of HMPs in digital mobility systems, and hence could replace established operators.

By taking the key findings as well as key issues derived from the scenario analysis into account several requirements which may be placed on future mobility service providers can be identified. The overarching requirement might be seen in the provision of services, which are accessible in a convenient manner. As found out by Accenture (2014), the research's results contain evidence for the increasing demand for convenience and reduction of mental effort to be put into buying decisions as well. Therefore, it may be crucial to place a high value on comfort and convenience in mobility service design. In addition, the discrepancy in the attractiveness of certain VAS seems to advocate for the importance of a true one-to-one customer management. In particular, the introduction of customer-specific CEM may be vital to balance this divergence in acceptance of proactive service offerings. Convenience and comfort could assume the shape of an integration of mobility (services) into everyday life activities as well. As the results demonstrate the existence of a cultural barrier towards ADV it may be claimed that—at least in the beginning—such vehicles should be utilised to a limited extent. The survey data proposes that ADV would rather take over assistive tasks such as the control on long-distance trips or to dissolve traffic jams. Hence, it ought to be considered to assign ADV with valet-parking type services. Nevertheless, as the trust in this technology increases the services should be gradually developed to full vehicle on-demand services.

7.2   Critical Review

Findings and results drawn from this research should be reviewed critically. It should be emphasised that the current stage of research in the field of digital mobility is mostly characterised by exploratory studies. Therefore, it ought to be avoided to apply too rigorous constraints in terms of definition of exact components, for instance. Business models as well as components of digital mobility discussed may be considered as a representation of feasible alternatives. Moreover, the likeliness of cooperation between data-driven and asset-driven has not been taken into account in this research.

Although several biases were considered in the research design, a complete elimination of every bias might be regarded as highly unlikely. Hence, the survey's findings should be reviewed critically. Firstly, the definition of key terminology had not been provided in the questionnaire distributed to the survey panel. Therefore, the confusion of terms or rather the unequal interpretation of terms between the author and the survey panel cannot be excluded. The strong tendency towards Internet companies observed in the survey data could be biased by the 'hype' made around feasible market entries of firms such as Google. In addition, companies that were selected the most in the subconscious survey in terms of capital investment shared the common characteristic of involvement in the development of ADV. Since various survey questions were centred around the usage and attractiveness of ADV, the occurrence of a halo effect on these questions may not be excluded. Besides, the analysis of results in general may be prone to misinterpretations.

Furthermore, in order to calculate the digital mobility split several assumptions had to be made, since the majority of services discussed are non-existent in the current state. Therefore, the results should be considered cautiously. Besides, the DMS calculation was carried out under the condition of treating Likert-type scales as metric scales which is not accepted by every scholar (cf. Bortz & Döring, 2006, 181 f.). Nevertheless, the developed DMS may be considered as valuable insight into the structure of future mobility markets from a customer's point of view. In addition, the DMS presented depicts the advocated service-based view from a planning perspective, as it is calculated on the basis of service bookings. A DMS portraying the operational side might be structured completely differently.

7.3   Limitations and Further Research

This research is subject to several limitations which provide opportunities for further research. First of all, the research's survey targeted the demand side of the mobility service industry. Therefore, future research could aim for studies from a supplier's point of view (i.e. mobility services providers) in order to contribute to the discussion. Secondly, as quantitative research was applied, a qualitative approach comprising in-depth customer and expert interviews could yield valuable insights as well. This might be especially beneficial with regards to the acceptance of CCM practices, in particular the willingness of usage of (non-mobility) VAS.

Furthermore, as this research solely focussed on Germany, it could be repeated in other countries. To contrast the findings future research might aim for countries with a lower level of uncertainty avoidance according to Hofstede Centre (2016) or a less car-centred culture.

Besides, the findings of this research could be utilised in order to conceptualise the digital transformation of currently laggard integrational companies such as public transport operators. This should include strategies to generate, process as well as provide sufficient user-specific data. In addition, the discussed DMS may be developed further, amongst others by introducing additional variables into the construct. In this regard, the inclusion of external information from previous studies concerning mobility behaviours such as Bartz (2015) or Henkel et al. (2015) might produce beneficial insights.

Further, future studies should take into account instrumental weak points discovered during this research. Firstly, the questionnaire survey should be repeated with data gathered under probabilistic sampling methods. Secondly, information provided to the survey participants could be optimised in terms of response support to reduce the previously discussed uninformed response bias. One example may be seen in the limitation of companies to the respective HMP business model. This could sharpen the results even further.

References

Accenture. (2014). *Customer 2020: Are you future-ready or reliving the past?: Ten years of Accenture research highlights real opportunities for providers to better meet customers' steadily rising expectations* (No. 14-6613). Accenture.

Bartz, F. M. (2015). *Mobilitätsbedürfnisse und ihre Satisfaktoren. Die Analyse von Mobilitätstypen im Rahmen eines internationalen Segmentierungsmodells* (Doctoral Thesis). University of Cologne, Cologne.

Bortz, J., & Döring, N. (2006). *Forschungsmethoden und Evaluation für Human- und Sozialwissenschaftler* (4th ed.). Springer Medizin Verlag.

Henkel, S., Tomczak, T., Henkel, S., & Hauner, C. (2015). *Mobilität aus Kundensicht.* Springer Fachmedien Wiesbaden.

Hofstede Centre. (2016). *What about Germany?* Retrieved August 10, 2021, from https://geert-hofstede.com/germany.html

Rammler, S., & Sauter-Servaes, T. (2013). *Innovative Mobilitätsdienstleistungen* (Arbeitspapier No. 274). Hans-Böckler-Stiftung.

The manufacturer's authorised representative in the EU is Springer
Nature Customer Service Centre GmbH, Europaplatz 3, 69115 Heidelberg,
Germany. If you have any concerns regarding our products, please
contact ProductSafety@springernature.com

Printed and bound by CPI Group (UK) Ltd, Croydon, CR0 4YY
29/04/2026
02099527-0004